Man's Impact on the Hydrological Cycle

in the United Kingdom

MAN'S IMPACT

ON THE HYDROLOGICAL CYCLE

IN THE UNITED KINGDOM

edited by

G. E. Hollis

ISBN cloth 0 86094 024 1
     paper 0 86094 018 7

published by Geo Abstracts Ltd.,
           University of East Anglia,
           Norwich NR4 7TJ,
           England

CONTENTS

continued overleaf

THE URBAN ENVIRONMENT

ACKNOWLEDGEMENTS

The idea for the seminar, from which this book grew, was
floated by Professor E.H. Brown. He and Professor P. Novak
chaired the meeting at the Institute of British Geographers'
Annual Conference in 1977. The experience of colleagues
at University College London was invaluable during the
early stages of the volume and the skill of the Carto-
graphic Unit and secretarial staff was much appreciated.

Permission to publish copyright material has been
granted by the Ordnance Survey, Elsevier Scientific
Publishing, the Journal of the Fisheries Research Board
of Canada, Dr. W. Junk B.V., Nordic Hydrology, and Dr. J.A.
Taylor.

The cover photographs are reproduced with the permission
of the Press Association and show (front) Severn valley
floods near Shrewsbury in December 1964 and (back) flooding
in central Bath in December 1960.

ACKNOWLEDGEMENTS

PREFACE

In 1974 the water industry in England and Wales was
reorganised into ten water authorities responsible for all
aspects of the hydrological cycle; the development of
water resources, water distribution and supply, the prevention of pollution, sewerage and sewage treatment, land
drainage and sea defence, and recreation and fisheries.
The role of the National Water Council is that of a con-
sultative and advisory body, with the ability to speak or
to act on behalf of all the water authorities where
appropriate.
    Consistent with their all-purpose role, the areas of
the water authorities are not based on administrative areas
but on river basins, the natural hydrological divisions of
the country. Water authorities have a good working knowledge of the natural hydrological cycle as it affects
rainfall, run-off and groundwater in their regions but more
research is needed to improve this knowledge in some
respects. However, to water authorities the advancement of
knowledge is not an end in itself, and for research to be
useful it must be communicated and understood. This volume
has ensured just that.
    This collection of papers is clearly directed towards
one theme, man's impact on the hydrological cycle in the
U.K, and it demonstrates how much the natural water cycle
is being constantly modified by human action. In some
cases these modifications are helpful to water authorities,
but all too often the effect of one area of human activity
can be to exacerbate problems experienced somewhere else.
For water authorities it is essential to understand and, as
far as possible, to anticipate this interaction.
    I am impressed by the food for thought these papers
provide for practical managers of the water cycle. Much of
the work described has been carried out in cooperation with
particular water authorities and some managers will be well
aware of these interactions from their own experience.
Another value of collecting the papers together is that it
draws the potential effect of human activities on the
hydrological cycle to the attention of a much wider
audience, particularly those engaged in forestry and
agriculture, urban development and flood protection. I
hope they will find this volume interesting and useful.

Lord Nugent of Guildford, PC
Chairman of the National Water Council
1973-1978

1

THE HYDROLOGICAL EFFECTS OF MAN'S ACTIVITY:
A U.K. PERSPECTIVE

G.E. Hollis

Department of Geography, University College London

ABSTRACT

Investigations of the hydrological effects of man's activity have increased significantly during the last decade. Studies of water quality effects have become especially important recently. There have been many international symposia but none have focussed exclusively on the U.K. Land use effects cannot be generalised for all climates so texts focussing on one part of the Earth are valuable.

INTRODUCTION

Interest in man's impact on the hydrological cycle has blossomed during the last decade. The seeds of this development were sown by the early experiments in the United States whilst, in Britain, Frank Law's (1956) work on the water balance of moorland and forest water gathering grounds at Stock's Reservoir was a major post-war stimulus. The effect of man's activity has been an element in the development of hydrology itself. The International Hydrological Decade (1965-74) pursued a project entitled 'The Influence of Man on the Hydrological Cycle' and this has been continued through International Hydrological Programme projects (IHP, 1975). Project 5 involves the investigation of the hydrological and ecological effects of man's activities and their assessment whilst Project 6 focuses on the hydrological and ecological aspects of water pollution. The effects of urbanization on hydrological regime and on quality of water, and the long-term prediction of ground-water regime, taking into account human activities, are Projects 7 and 8 respectively.
    The 1963 Water Resources Act in Britain 'led to the creation of river authorities and Water Resources Board, all of which had strong hydrological interests' (NERC,

1976b), and the more sweeping changes of the 1974 Water Act have strengthened and integrated the management of all aspects of the hydrological cycle. The Institute of Hydrology was established in 1968 with one of its main aims being 'to obtain and develop high quality catchment data' (McCulloch, 1969) for several experimental basins, some of which were aimed at an investigation of land use effects. All United Kingdom bodies with hydrological research interests were surveyed by NERC in 1965-70 and again in 1970-75 (NERC, 1975a). Smith (1976) showed, from NERC's data, that applied hydrological research increased by 310 per cent in those five years, the number of projects growing from 29 to 90. Interest in the subject continues to grow in the mid-1970s, for in a review of recent research, Walling (1977) says that 'closely associated with the growing interest in the interface of hydrological processes and other aspects of environmental systems and developing a theme which has become increasingly emphasised in recent years has been the considerable volume of literature concerning the influence of man on the hydrological cycle'.

THE AREA OF STUDY

The definition of the field of study entitled 'Man's Impact on the Hydrological Cycle' is impossible, for it encompasses all of the interactions between man's activities and the many facets of the land phase of the hydrological cycle. Until the 1970s studies of water quantity predominated but 'the diversification of the scope of physical hydrology to include quality' (Walling, 1977) has been especially marked in this field. The effects of land use changes, notably afforestation and urbanization, are major areas of enquiry. The implications of farming and forestry activities, e.g. drainage, tree-felling and fertilizer application, form another area of interest along with the storage, abstraction and return of water by urban communities. Inadvertent effects, such as changes in microclimate or downstream effects of a flood alleviation scheme are often seen as a distinct sub-section. The breadth of study poses major problems for both the analytical scientific hydrologist and the applied hydrologist concerned with prognostication. The former must design experiments and conduct sophisticated analytical programmes in order to discern the effect of man within the natural variation in the system. UNESCO (1972) said that 'present knowledge and experience of water resources can be combined to give some estimate of the hydrological consequences of major changes in land use but experimental studies on a watershed basis are a wise first stage in development'. The engineering hydrologist may have to take account of the impact of man's action on an event which may be so rare that it has not yet been observed under natural conditions. Costin and Dooge (1973) stressed the problems for the applied hydrologist by saying that 'with the rapidly accelerating pressure on the world's limited water resources and powerful developments in technology..., man's ability to achieve harmful ... results is increasing, and the safe margin of error is correspondingly diminished'.

The burgeoning of interest has produced a number of international volumes of research results. Most of them consist of collections of papers and there is generally a heavy concentration on United States studies. A significant early contribution was *Man's Role In Changing The Face Of The Earth* (Thomas, 1956) because it considered the whole of the natural environment as well as sociological, political, cultural and economic issues. A similar perspective has been adopted more recently by Detwyler (1971), whilst Ryabchikov (1975) examined the earth's geosphere as a physical system before considering man's effect from a socialist viewpoint. More narrowly focused texts have included the Ackermann et al. (1973) review of the problems and environmental effects of man-made lakes, the UNESCO (1975) compendium of work on the hydrology of marsh-ridden areas which included a section on the hydrological and meteorological effects of reclamation techniques, and the symposium on urban climates and building climatology (WMO, 1970a and b). The hydrological problems of forest areas were reviewed in Sopper and Lull (1965), whilst soil erosion and sediment yields from urban and rural areas are considered by IAHS (1974) *The effects of man on the interface of the hydrological cycle with the physical environment*. More strictly within the confines of hydrology, Moore and Morgan (1969) edited a seminal volume on the effects of watershed changes on streamflow based wholly on American research. The book has sections on urban and rural areas including one paper in each section on water quality. Whipple (1975) examined the quantity and quality of urban runoff with a special emphasis on its management in the United States context, whilst Helliwell (1978) looked at urban storm drainage. Two other volumes have looked specifically at urban hydrology: the UNESCO-IAHS Symposium in Amsterdam in October 1977 (IAHS, 1977) looked at the effects of urbanization and industrialization on the hydrological regime and on water quality as well as water policy as a factor in urbanization. The UNESCO (1974) compendium on the hydrological effects of urbanization summarized international research findings, set out a series of priorities for international action and presented both national and special topic case studies. There have been two major reviews of man's impact on the hydrological cycle stemming from the IHD. The IHD Working Group on the influence of man on the hydrological cycle (UNESCO, 1972) collected outstanding studies of the hydrological effects of major land use changes and found a need for positive land use policies under the headings of forest lands, grasslands, arable lands, irrigation and salinity, swamp drainage, urbanization and water pollution, landslides and road construction and the location of hazardous areas in large catchments. This was essentially a review of the impact of these changes on the hydrological cycle and a compendium of scientific research work investigating the changes. FAO (1973) complemented this report with a second, using a different approach to the subject. They argued that 'man's efforts to control the world's water cycle invariably involve factors other than hydrology and engineering: ecological, sociological, economic, cultural and political consid- erations and forces'.

THE UNITED KINGDOM CONTEXT

There have been a number of British contributions to these international symposia but few volumes have focused exclusively on the U.K. CIRIA (1974) organized a symposium on rainfall, runoff and surface water drainage of urban catchments but the emphasis was very largely on sewer design and evaluation. The Institution of Civil Engineers' (1975b) review of contemporary engineering hydrology scarcely mentioned human influences on the hydrological cycle. The seminar to mark the opening of the Maclean Building of the Institute of Hydrology included a paper by Pereira (1973a) which examined the increasing importance of comprehensive measurements of catchments where man's activities are important. He argued that the hydrological effects of advected energy, the water and energy balance of a permeable catchment, and the hydrological problems of land use changes overseas are the main scientific issues remaining to be tackled.

The present book, which developed from a symposium at the Institute of British Geographers' Annual Conference at Newcastle-upon-Tyne in 1977, presents a collection of papers describing current work in the U.K. in the field of man's impact on the hydrological cycle. The bulk of the papers cover scientific investigations of particular aspects of man's activity. The aim of the book is to collect together the results of these current scientific investi- gations for the benefit of scientists and managers. Most chapters discuss the management implications of the research findings and the concluding chapter sets the results in the context of catchment control as exercised by the regional water authorities in England and Wales. The regional concen- tration on the United Kingdom was chosen to allow a fairly comprehensive coverage of current research in this country, to concentrate research results for water managers in the U.K., and to present a range of studies from one climatic zone, for as UNESCO (1972) said, 'the hydrological effects of land use cannot be generalised for all climates'. The division into rural and urban environments is perhaps a little arbitrary but reflects two rather distinct areas of human activity and follows the pattern set by Moore and Morgan (1969) and IAHS (1974). A volume of this type cannot hope to be comprehensive, for there were 25 projects dealing specifically with human activity in the NERC (1975a) Census and many investigations, including some reported here, have been started since that 1975 survey. Perhaps the most important omission is a contribution devoted to the nitrate content of surface and underground waters. Foster and Crease (1974), for example, examined the chalk aquifer in East Yorkshire and found high levels of nitrate in water which was probably more than ten years old. An example of the nitrate pollution of river water is Slack's

(1977) examination of the Essex rivers where very high
concentrations of nitrate were the result of water draining from agricultural land.

THE PROSPECT

Future research on man's effect on the hydrological cycle
is likely to follow the current movement towards a systems
view of the whole hydrological cycle, the concentration
upon water quality as well as the water quantity, and a
reliance upon numerical modelling. The NERC Working Party
on Hydrology (NERC, 1976b) identified the effects of soil
moisture and vegetation changes on actual and potential
evapotranspiration and the ongoing improvement in numerical
models to describe catchment behaviour as long-term
strategic research topics. The influence of urbanization
and other land uses and drainage on evaporation and runoff
was seen as a more specific or immediate research need.
Lowing's (1977) review of urban hydrological modelling and
catchment research in the U.K. described a number of
active research projects aimed at gathering reliable data
on quantity and quality aspects of urban runoff. He also
found that 'urban hydrological modelling in the U.K.
continues to be geared primarily to the improvement of
sewer design methods'. The Flood Studies Report (NERC,
1975b) incorporated an urbanization factor into their
predictive equations for flood flows. A subsequent seminar
(NERC, 1977) showed that more research was needed because
'the present FSR methods ... should be limited to catchments where the urban fraction does not exceed 25 per cent
and ought not to be used to predict the effects of
increasing urbanization'. At the international scale, an
IHP workshop (IHP, 1978) concluded that in urban areas
'further effort is required to improve both the data base
and performance reliability of modelling techniques,
particularly in water quality modelling, in order to
increase their acceptability'. The future therefore holds
a prospect of continuing challenges for scientific
hydrologists in the areas of measuring, modelling and
predicting the effects of human activity on the hydrological
cycle in the U.K. This volume represents a cross-section
of current U.K. work in a field which is moving forward
rapidly but widening in scope as it does so.

THE RURAL ENVIRONMENT

2

FIELD UNDER-DRAINAGE AND THE
HYDROLOGICAL CYCLE

F.H.W. Green

Department of Agricultural Science, University of Oxford

ABSTRACT

*Field under-drainage is usually undertaken with earthen-
ware or plastic pipes but secondary treatment by moling or
sub-soiling is often employed as well. Flash flooding is
reduced by under-drainage whilst the total volume of river
discharge may increase. Data from two catchments illustrates these effects and the national extent of under-
drainage is charted. The effects of under-drainage on
water quality are illustrated by a discussion of nitrate
levels in rivers and well water.*

INTRODUCTION

Irrigation and drainage are both undertaken to improve
agricultural potential by keeping the soil in more nearly
optimum conditions at all seasons of the year. For
example, in East Anglia, where topography and temperature
conditions are generally the best in the U.K. for arable
agriculture, the water balance is not ideal. There is
often a shortage of water in the summer half of the year,
when evaporation and transpiration exceed rainfall, and
yet ploughing and other field operations are often
hampered by waterlogged soil in the winter half-year. So
although irrigation is extensively practised in the summer,
there has been more land drainage than in any other part
of the country.

The two types of land drainage are the improvement of
watercourses and field drainage, which correspond roughly
with the terms surface drainage and under-drainage.
Arterial drainage usually refers to the regulation of major
watercourses, by straightening, deepening, embanking, and
sometimes pumping, as exemplified in the English Fenland.

Surface drainage is almost entirely undertaken in the
U.K. by statutory bodies. It is not usually undertaken

solely for agricultural purposes. Other aids to surface run-off are often excluded from the term arterial drainage. The medieval ridge-and-furrow system, however it originated, was perpetuated because the furrows facilitated run-off of surplus water into streams, which were in turn kept clear. Where streams were lacking, ditches were excavated. Nowadays, farmers can apply for grants to improve ditches in the same way as for grants to undertake field under-drainage.

Under-drainage of fields, which is the main subject of this paper, includes all methods of facilitating the movement of water through the soil, especially downwards from the upper layers. Today the usual procedure is to lay earthenware or plastic pipes, in lines, at a depth of about 100-120 cm. In some conditions such tile-draining is sufficient in itself to carry off water, but in heavy soils it is desirable to carry out secondary treatment to facilitate the passage of water from the soil above down into the tile drains. Secondary treatment includes moling and sub-soiling; the first of these consists in forming unlined channels above and approximately at right angles to the tile-drains, and the second consists in effect of deep ploughing to break up the soil, the effect of the two being generally similar. Another aid to the passage of water into the tile-drains is the use of a permeable back-fill such as gravel above the drains.

The effective life of unlined mole-drains and of sub-soiling is, not surprisingly, much less than that of the underlying tile-drains and they commonly ought to be renewed within considerably less than a decade. Moling is easier to undertake when the surface is firm and the ground is dry, and so it predominates, not so much where it is most needed, but where the conditions most frequently occur for it to be undertaken (Figures 1 and 2).

HYDROLOGICAL EFFECTS

The hydrological effects of field under-drainage depend upon the type of under-drainage adopted and soil characteristics, particularly interflow properties. If satisfactory outflow is provided for the drains there will be a lowering of the average level of the water-table in the soil.

Figure 3 illustrates the effects of tile and mole drainage during the winter at a site in Suffolk. It is one of a series of some 300 sites planned jointly by the Soil Survey of England and Wales and the Ministry of Agriculture for the Soil Water Regime Study.

It is less easy to generalise about other hydrological effects because soils differ in conductivity, clays in particular behaving differently in their wet and dry states. The zone subject to interflow can vary and under-drainage varies in efficiency, the amount of secondary treatment being a relevant factor. Particular account must be taken of the initial state of the soil before heavy rain.

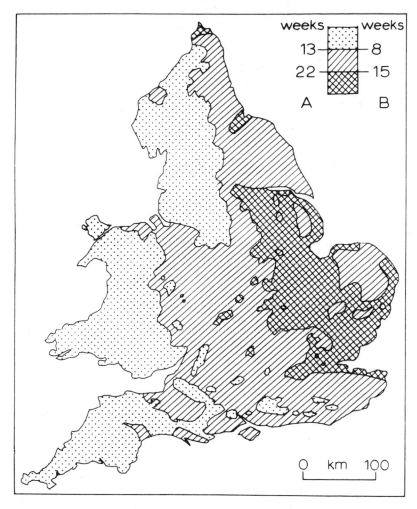

weeks   weeks

13 — 8
22 — 15

A   B

0   km   100

Figure 1. Average length of moling season. (A) Ideal,
when ground is hard or cracked. (B) Maximum, when
ground surface is firm. Source: Provisional map by
the Field Drainage Experimental Unit.

Assuming that under-drainage is achieving a lowering of
the water-level in the soil and a minimization of the
periods of saturation, three effects normally result.
First, the frequency and amount of surface run-off are
reduced and, since surface run-off is usually responsible
for the highest peaks of discharge, the frequency of
'flash-flooding' is reduced. Second, the volume of discharge is increased, but the peaks are flattened as
compared with those created by surface run-off. Third, as
the amount of under-drainage in a small catchment is
increased, the heights of the peaks tend again to rise
slightly but the recession is more rapid than before.

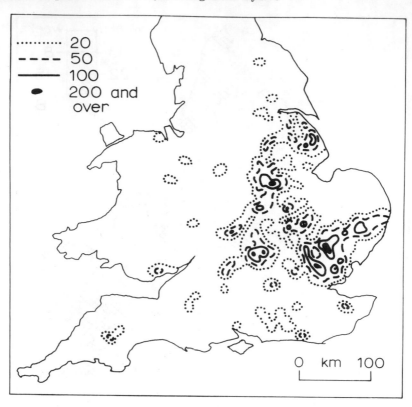

Figure 2.   Area moled (ha per km²) in 1972-73.   Source:
   Green (1976).

Various specific small-scale and short-period experiments have been undertaken to verify the above assertions.
An attempt has also been made to follow the sequence of
events over the years in two catchments each of about
100 km² in the east of England (Green, 1975).

About 7% of the catchment area of the Harper's Brook
in Northamptonshire had been under-drained between 1940
and 1974 (mostly between 1950 and 1960 and since 1968).
In the case of the Bury Brook in neighbouring Cambridge-
shire over 30% of the total area had been underdrained in
the same period. Gauging weirs installed at the lower ends
of the two catchments indicate that the number of peaks
above a conveniently chosen threshold has increased prop-
ortionately more in the Bury Brook than in the Harper's
Brook. However, the number of peaks has increased in both
catchments, and this increase coincides with a known
increase in short period heavy falls of rain since about
1967. More work needs to be done to determine what
proportions of the noted effects can be attributed to field
under-drainage and to the incidence of rainfall.

In modelling the hydrology of catchments, it is now

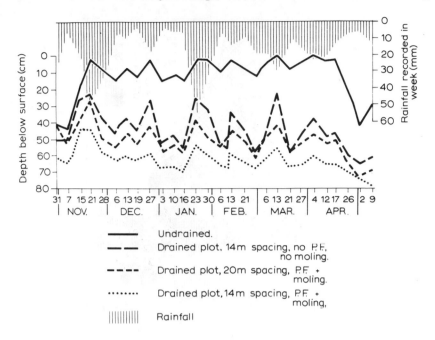

Figure 3. The effect of tile and mole draining on soil
water levels, winter 1974-75. Source: Field drainage
Experimental Unit.

realised that the effects of under-draining must be taken
into account. The Institute of Hydrology has recognised
that its lowland River Ray experimental catchment area in
Buckinghamshire has been receiving under-drainage similar
in intensity to the Bury Brook, especially during the last
15 years. The outflow from one large tile-and-mole
drained field within the River Ray catchment is now to be
monitored.

The extent of under-drainage in England and Wales since
1940 has been mapped (Green, 1973 and 1976) and it has been
shown that currently more than 2% of the total area of
several eastern counties, is receiving under-drained
treatment each year, although it must be noted that some
of this is the renewal of secondary treatment. Over 15%
of the total area of Essex, and over 30% of its arable
area, was under-drained between 1940 and 1972. For
the whole of England and Wales, 0.72 percent of the total
area was under-drained in the year 1973/74 (Figure 4). By
contrast, the percentage of the total area of France under-
drained in 1974 and 1975 was 0.08 (Figure 5). There was
of course considerable under-drainage undertaken in the
middle of the nineteenth century, but the data quoted by
Phillips and Clout (1970) show that much less field
drainage was undertaken in the nineteenth century in France
than in Britain. Consequently under-drainage is a much
more important factor influencing the hydrological cycle
in England and Wales than in France.

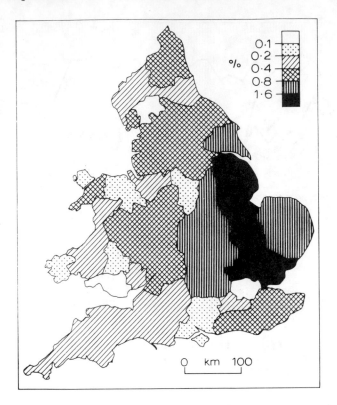

%
0.1
0.2
0.4
0.8
1.6

0 km 100

Figure 4.   Percentage of the total area under-drained in
1973-74 for England and Wales.

Some field drainage has of course been facilitated,
without under-drainage, by construction, deepening, and
widening of ditches and by embanking and pumping;   France
and Britain are probably more nearly equal in this respect.
However it is worth noting that tile-drainage directly
leads to increase of surface drainage effort.   It has to
be ensured for instance that the bottom of ditches is
below the level of the tile-drain outflows.   The process
of under-drainage therefore leads to considerable alteration of the network of surface watercourses.   In some
cases the catchment boundaries have been altered, for
instance in an area of subdued topography shallow ditches
which formerly led to one river have had their functions
replaced by ditches deepened to accommodate tile-drain
outflow which lead to a different river.

EFFECT ON WATER QUALITY

Hydrology is concerned with water quality as well as water
quantity.   Change in nitrate content of stream water and
well water illustrates the qualitative effects of field

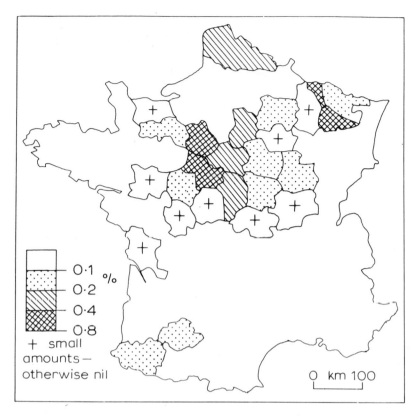

Figure 5.   Percentage of the total area under-drained in
     1974 and 1975 for France.

under-drainage.   An increase has been widely noted and
parallels the increased application of nitrogenous
fertilisers, notably since the mid-sixties (Figure 6 and
and 7).   This implies a waste of fertilisers although
farmers do not consciously over-fertilise.   There is
circumstantial evidence that fertilisers have been
applied inadvertently at a time and place favouring the
throughflow of a significant proportion.   Certainly
there always tend to be peaks of nitrogen content
in watercourses in spring when the outflow of water
decreases more than its solute content and in autumn when
outflow recommences and brings with it surplus nitrogen
accumulated during the summer.   This was especially so in
the wet period following the severe drought of 1976 when
unacceptable nitrate levels were recorded in some watercourses.   The amount of nitrates released through drains
is dependent in part upon biochemical processes within the
soil, which are in turn dependent upon the physical
treatment of the soil, including drainage.   Although the
contribution of agricultural fertilisers to the nitrate
content of watercourses has been increasing, it is still

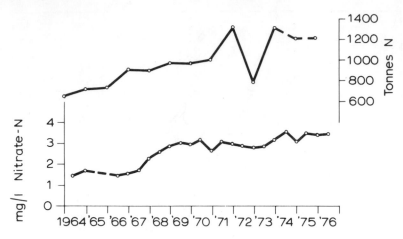

Figure 6.   The concentration of nitrogen in Loch Leven
        (below) and the tonnage of nitrogenous fertiliser (N
        equivalent) subsidised in the County of Kinross
        (above).   Note: The peaks were caused by the
        impending reduction of the subsidy after 1971-72 and
        its impending withdrawal after 1973-74.   Source:
        Holden (1976).

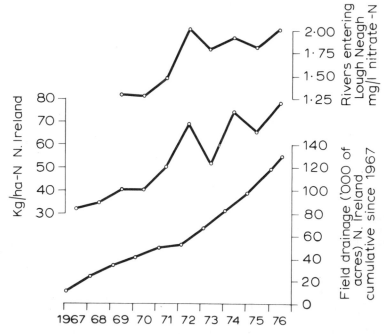

Figure 7.   Nitrogen concentration in rivers entering
        Lough Neagh (top), the application rate of nitrogenous
        fertiliser (N Equivalent) in N. Ireland (middle) and
        field drainage in N. Ireland (bottom).   Source:
        Smith (1976) and Department of Agriculture, N.I.

less in general than the contribution from sewage discharge and other sources. However increases in the rate
of throughflow of water because of under-drainage can
increase the levels, especially maximum levels, of
nitrogen in stream and well water.

ACKNOWLEDGEMENTS

I am indebted to many people, but particularly to
the following, not mentioned in the text or references:
B.D. Trafford of the Field Drainage Experimental Unit, and
to J.L. Devillers of C.T.G.R.E.F. (Ministry of Agriculture)
Antony, France.

3

SEASONAL CHANGES IN MICROTOPOGRAPHY AND SURFACE DEPRESSION STORAGE OF ARABLE SOILS

I. Reid

Department of Geography, Birkbeck College, London

ABSTRACT

Autumn agriculture leaves ploughland to the ravages of winter weather. The result is a progressive reduction of surface roughness with increasing soil settlement, the rate of change being functionally related to soil type and field management. Surface depression storage is minimal at a time when young plants offer little surface protection, leaving the soil susceptible to high recurrence-interval spring rainfalls. This seasonal decline in surface roughness is followed for two English soils using a series of contour maps and isometric diagrams. While a rendzina suffers insignificantly from one winter's weather, a multivariate analysis relates the considerable settlement of a clay soil to the number of wetting-drying cycles, rainfall intensity-duration, and the magnitude of soil micro-relief. This continuous readjustment of the microtopography determines a time-dependence for surface depression storage capacity of ploughland which may be as much as halved by winter weather.

INTRODUCTION

This is a description of the way the ploughman secures an annual increase in surface roughness and the way the weather destroys it. It is an account of the small-scale changes that occur in ploughland surface morphology following primary cultivation of arable fields. In general, it involves the breakdown of clods left by disc harrowing into aggregates and individual particles which, upon detachment, move freely into intervening hollows. The overall effect is not only a natural preparation of the seedbed, so especially important to farmers of claylands, but also a progressive reduction in the water storage capacity of the soil surface.

Primary cultivation provides an exaggerated surface
depression storage that is, for many soils, eroded by
winter weather.

The implications for catchment hydrology and for problems of general water management are underlined by the
fact that surface roughness is constantly diminishing
during a season of general water surplus, and at a time
when the soil surface lacks any significant protection
from plant cover. The hydraulic conductivity of the soil,
its ability to handle infiltering water, may well be at
its annual maximum (Reid, 1975), but infiltration capacity
is at a minimum (Horton, 1936). The situation is exacerbated should a plough pan of low permeability effectively
reduce the total short-term storage capacity of the surface soil layer, or should infiltration capacity be dramatically reduced under prolonged or intense rainfalls
where soils are liable to rapid slaking. For all these
conditions, rainfall excess is highly likely, and the
retention of rain water at site of ground contact depends
largely upon the capacity of contemporary surface
depression storage. As winter progresses and the hollows
fill with material from upstanding clods, so the probability of a cumulative rainfall excess larger than concurrent surface depression storage capacity increases, with
the attendant risks of soil erosion by surface runoff.

Surface depression storage as a *time-dependent*
variable is not accounted for in present empirical or simulation models of catchment rainfall-runoff response. Most
hydrological treatises give it a respectful, but cursory
glance (for example, Ward, 1975; Wisler and Brater, 1959),
and those research papers that consider its effect (Langford
and Turner, 1973; Smith and Woolhiser, 1971;
Swartzendruber and Hillel, 1975) treat it conveniently as a
constant storage capacity. This can only reflect the
difficulty of characterizing so geometrically complex a
variable, and one that changes continually.

The actual role of surface depression storage in
reducing problems of surface runoff and promoting infiltration is indicated in the results of plot rainfall-runoff
experiments for contour-ploughed slopes on the one hand,
and 'straight-row' ploughing on the other. Ellison (1974b)
shows a 50 per cent reduction in the median velocity of
overland flow on contour-ploughed 9 degree slopes, while
Glymph and Holtan (1969) report a 28 per cent greater
annual water retention for a catchment subject to contour-
ploughing compared with a neighbouring catchment under
'straight-row' cultivation.

THE SOILS

The experiment reported here follows the changes that occur
in the surface configuration of two soils, one of which has
been subject to two agricultural treatments. The soils
were collected after primary cultivation had been completed.
In each case, the last operation consisted of disc-
harrowing. One of the soils (Alnwich Farm, Figure 1),
taken from the Vale of Aylesbury, is mapped by Avery (1964)

Table 1. Soil index properties

	Particle-size fraction (%)				Dry Density*	Org-C	pH
	< 2 μ (clay)	2-63μ (silt)	63-2000μ (sand)	> 2000 μ (gravel)	(g/cm³)	(%)	
Alnwick (East) GRL1	51	35	8	5	0.90	3.81	8.0
Alnwick (West) GRL2	62	33	4	1	0.95	3.92	8.0
Hale (GRL3)	28	56	13	3	1.18	2.47	9.0

* at time of sampling

as Wicken Series and is derived from Gault Clay. A small
amount of flint gravel (see Table 1) reflects a certain
degree of superficial re-working. The other soil (Hale,
Figure 1) is mapped (Avery, 1964) as a rendzina of the
Icknield Series, and was taken from a scarp valley in the
Chalk cuesta near Wendover.

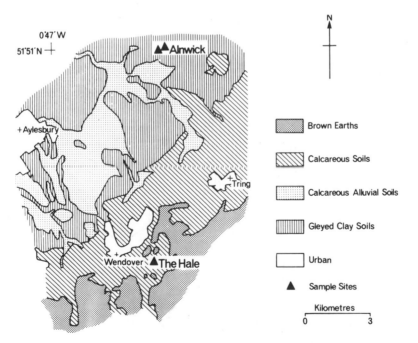

Figure 1. Index map of soils and sites. Soil boundaries
    extracted from Avery (1964).

Man's impact on the hydrological cycle

The Vale of Aylesbury soil classifies texturally as a clay. Two sample sites were located. The first (designated GRL1) was under plough for the first time in the farmer's memory. The second (GRL2), some 250 metres west of GRL1, was taken from a part of the field which had been ploughed for 5 successive seasons. Both sites had been old pasture before cultivation. The rendzina (GRL3) is a silty-clay loam with an abundance of calcium carbonate and had been under cultivation for some years.

EXPERIMENTAL DESIGN

At each site a specially constructed lysimeter was filled with soil from the plough-layer, paying particular attention to the need to maintain the same degree of disturbance as the in situ soil, and especial heed to replication of the surface character of the site zone. The surface area of each lysimeter is $0.24m^2$.

The soils were collected at the end of September 1973 and transported to the College roof laboratory. Here all the relevant hydrometeorological variables are recorded. Exposure was standardized. The lysimeters were inclined at 7 degrees to facilitate drainage and each was orientated to face south. The soils were then left to the weather, the only treatment being a careful clipping of any vegetative growth.

Measurements of the soil surface began on 22nd October and continued at 4-weekly intervals throughout the winter 1973/74 until May. For each lysimeter the spot-heights at

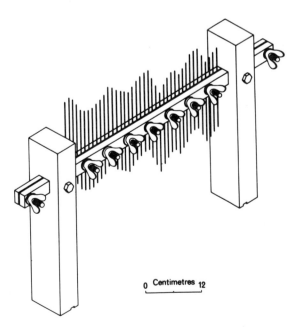

0 Centimetres 12

Figure 2.   Micro-relief meter.

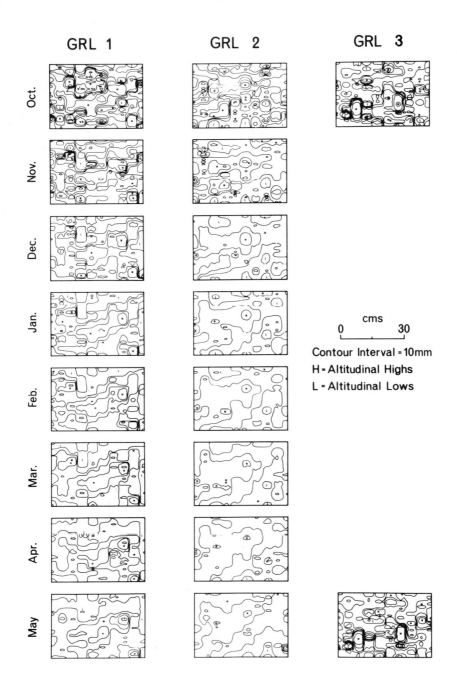

Figure 3. Soil surface contour maps.

GRL 1

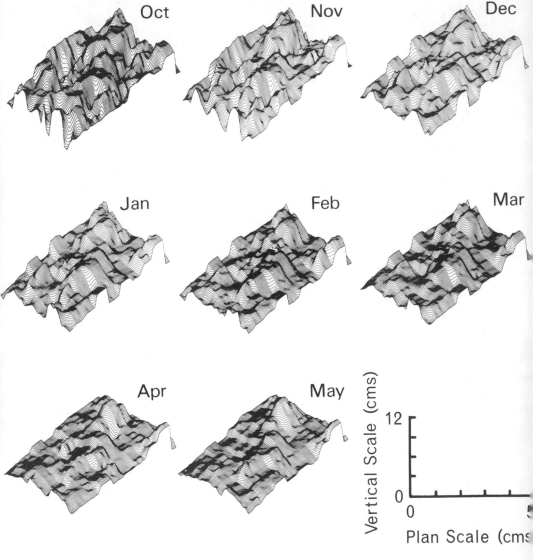

Figure 4A. (above) Isometric projections of surface
    configuration of newly ploughed old pasture clay soil.

Figure 4B. (opposite, top) Isometric projections of
    surface configuration of 5-year ploughed old pasture
    clay soil (for scale see Figure 4A)

Figure 4C. (opposite, bottom) Isometric projections of
    surface configuration of ploughed rendzina (for scale
    see Figure 4A)

GRL 2

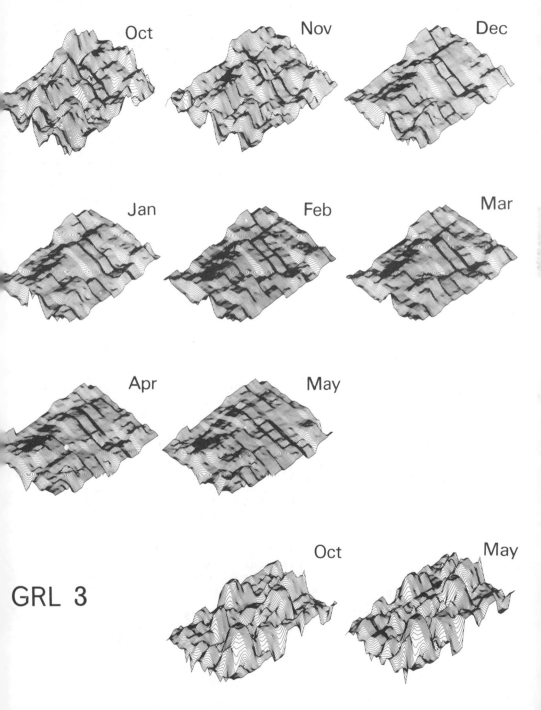

Oct

Nov

Dec

Jan

Feb

Mar

Apr

May

GRL 3

Oct

May

25

480 co-ordinate points were determined to the nearest
millimetre using a micro-relief meter (Figure 2) similar
to, but much smaller than those used by Kuipers (1957),
Kuipers and Van Oewerkerke (1963), Burwell, Allmaras and
Amemiya (1963), Wilton (1964), and Steinhardt and Trafford
(1974). In total, some 11520 spot-heights were recorded.
In order to allow for boundary conditions at the lysimeter
wall, a buffer zone of 5 cm width was selected, and the
spot-heights falling within this zone excluded from any
subsequent calculation. As a precaution, the immediate
roof area of the lysimeters was kept continually swept in
order to detect losses by outsplash, but none occurred.

RESULTS

Contours were interpolated using SYMAP (Lab for Computer
Graphics, 1975a) and checked stereoscopically with perpen-
dicular photographs taken on every sampling occasion. The
outcome is a detailed relief map of each soil surface as
it changes from late autumn to spring(Figure 3). These
are rendered more readily assimilable in the isometric
projections (Lab for Computer Graphics, 1975b) of
Figure 4A, B and C.
     Most noticeable in the contour maps or isometric
diagrams is that, while all 3 lysimeters start the winter
season with considerable surface roughness, the clay soils
are reduced to an undulating plain while the rendzina of
GRL3 remains largely untouched in its gross surface micromorphology. In fact, continuous observation of the rendzina from week to week revealed only a very gradual
emergence in bas-relief of the insoluble larger particles
as the surface fines were etched or dissolved.
     The two clay soils show differences in behaviour that
would not be expected (Greenland et al., 1975) merely from
the analysis of such colloidal binding properties as clay
content and organic matter (Table 1). Indeed, without any
significant differences between soils, the old pastureland,
newly ploughed, undoubtedly possesses that ill-defined
quality of soil that has for long been under grass, making
it cloddy (Russell, 1971; Low, 1972; Low, 1973; Low and
Stuart, 1974), and so giving it an initially greater range
of microrelief. The October surface roughness of GRL2
(ploughed on 5 successive years), given by the standard
deviation of the spot-height altitudes, is only approached
by GRL1 (the newly ploughed pasture soil) in December, two
months later.
     It takes approximately 3 months for GRL2 to lose most
of its micro-relief, thereafter being subject only to
minor adjustments in relative soil elevation (Figure 5);
but significant changes in the relief of the old pasture
soil (GRL1) occur as late as April, some 7 months after
primary cultivation. Five successive years of ploughing
has not only reduced initial relief of this Wicken Series
soil but permits more rapid attainment of a post-
cultivation soil plane.

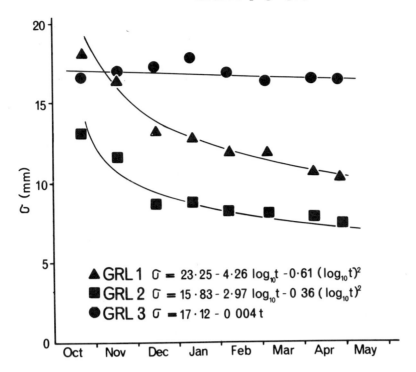

Figure 5. Soil surface roughness given by standard
deviation of spot-heights (σ) as a function of time.

PROCESSES

From the time of the first detailed measurements of surface
micromorphology on 22nd October until the last of 6th May,
205 mm of rain were recorded over 76 raindays by the roof
laboratory rain-gauge. January/February was the wettest
4-weekly interval and March/April the driest, though
November experienced the highest rainfall intensities.
Solar radiation for the sampling period totalled 37.2 W/cm^2.
Air temperature records reveal only 2 screen frosts in the
entire 7 months, a reflection not only of the extremely
mild 1973/74 winter but also of the very real 'heat island'
effect of urban fabric and artificial heating (Chandler,
1965). In fact, the considerable man-induced modification
of weather experienced in central London is extremely useful in two directions. In the first instance, the rates of
clod breakdown are probably representative of the minimum
to be expected under field conditions and reflect the
slowest modification of the soil surface by weather.
Secondly, notwithstanding mild winter conditions, the rapid
breakdown of the clay soils is ample indication of the
efficiency of the single process largely responsible for
disintegration of clods into individual aggregates.
    It may be safe to assume that frost, the subject of so
much debate concerning aggregate integrity (Benoit, 1973;

Bisal and Nielsen, 1964; Richardson, 1976), does not play a part in clod destruction in this experiment. Two other mechanisms have been considered important, rainsplash (Ellison, 1947a; Hudson, 1957; Rose, 1960; Young and Wiersma, 1973), and the shear forces developed through clay lattice expansion and contraction upon wetting and drying (Quirk and Panabokke, 1962).

In order to assess the combined effect of these variables and to allow for the dominant role played by the opportunity of dislodged aggregates to move either under gravity alone or with the encouragement of an applied stress, a multiple regression analysis was undertaken. The actual movement of material from high points into the depressions is represented by the change from month to month in the standard deviation of spot-heights, itself a statistical measure of the microtopographical range. This is regressed upon the contemporary opportunity for movement represented by the standard deviation of spot-heights, the number of rainfalls as a measure of wetting-drying cycles, and the duration of rainfall at intensities equal to or greater than 2 mm/hr, a threshold found to give the highest co-variation. Using these variables, the multiple correlation coefficient for the newly ploughed pasture soil (GRL1) is 0.958, significant at the 5 per cent level. The coefficient for the 5-year ploughed clay soil (GRL2) is not so good and stands at 0.870, not reaching the 5 per cent level. The partial correlation and regression coefficients are given in Table 2.

Table 2. Multiple correlation and regression of soil settlement (ΔS) on antecedent settlement opportunity (S_O), number of rainfalls (R_f), and rainfall intensity-duration (R_i)

			Multiple correlation	Partial correlation $r_{01.23}$ $r_{02.13}$ $r_{03.12}$		
Alnwick (East) GRL1	$\Delta S_O =$ 0.082-0.157S_O+0.119R_f-0.538R_i		0.958	-0.76	0.89	-0.89
Alnwick (West) GRL2	$\Delta S_O =$ 1.206-0.283S_O+0.073R_f-0.304R_i		0.870	-0.70	0.60	-0.56

One thing that arises clearly from the multivariate analysis is that the additive effect of rainfall intensity-duration is small. This is not surprising considering the size of the aggregates (modal b-axis, 6 mm) and the low intensities of British rainfall (Hudson, 1971), the maximum for the experimental period being 9.2 mm/hr. But the diminutive roles of both rainsplash and frost-action expose the significance of intra- and inter-aggregate stresses associated with water sorption-desorption. The progressive breakdown of the Wicken Series soils under these experimental conditions is almost entirely due to this process. Differences in the rate of degradation between newly ploughed and 5-year ploughed soil reflect a change in integrity of the old pastureland clods as they become subject to successive years as arable. Interestingly, the rendzina with lower clay content and the freely available cementing agent of calcium carbonate suffers insignificantly from one winter's weather.

Table 3.   Surface depression storage

Soil		Surface Depression Storage		
		October (mm)	May (mm)	Ratio October/May
Alnwick (East)	GRL1	6.91	3.43	2.01
Alnwick (West)	GRL2	5.20	3.06	1.70
Hale	GRL3	5.64	5.63	1.00

The consequence for surface depression storage of this seasonal adjustment in soil microtopography were estimated by careful flooding of scaled models reconstructed from the spot-height data. As expected, the rendzina retained its initial storage capacity. The clay soil, on the other hand, lost by May 50% (GRL1) and 41% (GRL2) of the exaggerated storage capacity conferred by autumn tillage (Table 3). Since it is more convenient to measure surface roughness than to estimate actual surface depression storage, it is more than useful that the two variables are highly correlated ($r = 0.966$, $D_S = 0.36 + 0.34S_O$ where surface depression storage, D_S, and surface roughness, S_O, given by the standard deviation of spot-heights, are both expressed in mm).

CONCLUSIONS

A quadratic relationship is demonstrated for post-cultivation settlement of an English clay soil. Real differences in behaviour of the same soil, either newly ploughed or following the cultivations of successive years, reflect progressive loss of cloddiness and an increase in susceptibility to seasonal weathering processes. The two curves (Figure 5),may be members of a family of settlement curves characteristic of this particular soil. The Icknield rendzina suffers little under winter weather, only showing slight rounding of the clods and an etching of its surface.
     In the case of the rendzina, surface depression storage may be regarded as a seasonal constant when modelling overland flow. The clay soil, on the other hand, presents a continuously changing surface geometry which moves more or less rapidly to a plane depending upon its history of

cultivation. The implications for water management problems in catchments dominated by clay soils and old ploughland are that the soils will more quickly become incompetent to deal with large recurrence-interval rainfalls as winter progresses and settlement reduces depression storage capacity.

ACKNOWLEDGEMENTS

Thanks are due to Christopher Hawkins who not only constructed the relief meter and lysimeters but also assisted in collection of the soils. Dr. L. Frostick gave up a lot of time in connection with the production of the isometric diagrams. For this help I am especially grateful.

4

AGRICULTURAL CONSEQUENCES OF
GROUNDWATER DEVELOPMENT IN ENGLAND

D.S.H. Drennan

Department of Agricultural Botany, University of Reading

ABSTRACT

Several schemes for the conjunctive use of groundwater and surface water are underway in England. Their aim is to augment summer riverflows by pumping groundwater into streams. Where groundwater development affects existing farm water supplies, Water Authorities are obliged to make good any detriment. Lowering of water tables by pumping can drain wetland thereby benefiting farmers but threatening eco-logically important wet habitats. Capillary rise is often regarded as a benefit to the farmer during summer droughts but it is shown to be of rather limited importance. The careful siting of wells and the monitoring of sensitive areas can obviate negative effects on agriculture and ecology.

INTRODUCTION

There are many parts of England where the potable ground-water lying within reasonable pumping depths in chalk, lime-stone and sandstone aquifers may be the only uncommitted water resource left for further development of domestic and industrial supplies. Groundwater development has many attractions for the water engineer. If a decision is made to use it, the rate of its development can be varied to meet the rate of demand. Surface storage schemes tend to create block supplies which tend to be in excess of requirements for some time until they are fully committed. Groundwater can also be used more flexibly. The most efficient and least costly wells can be used more than the others. A network of wells and their ancillary pipelines has a low total land requirement after the installation stages and can have little visual intrusion on the landscape if proper planning and design of installations is carried out.
Whilst most existing use of groundwater has tended to

build up in a piece-meal fashion over many years it is now
realised that a sensible conjunctive use of surface storage
and groundwater can often provide the best total management
policy of the water resources of a substantial area. The
normal base flow discharge of a catchment is a reflection
of the seasonal variation in groundwater levels and the
changing volume of the aquifer discharging at various times
(Figure 1). The addition of pumped groundwater to rivers

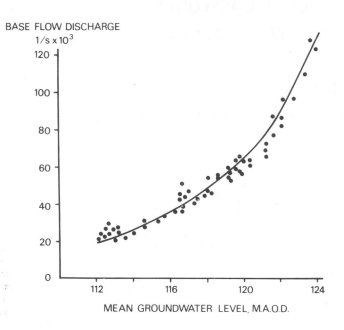

Figure 1. Correlation between the baseflow of the River
Lambourn at Shaw and the mean groundwater level in its
catchment area. (Source: Thames Water Authority).

to maintain their base flow rates above those amounts of
water discharging normally from springs, seepages and bed
transfer at periods of low summer flow is an important
feature of several current conjunctive use schemes in the
Thames, Anglian and Severn-Trent Water Authority areas.
This not only provides a larger amount of good quality water
at intake sites on such rivers but also has important amenity
and ecological value to substantial parts of river catchments
particularly in times of drought. The effect of pumping in
this way however results in a greater depletion of groundwater levels than would normally be the case leaving an
additional amount of storage to be made good during the next
recharge period. In this way the normal seasonal base flow
hydrograph of a river may be flattened out with smaller
excess flows over the late winter to early spring period and
larger base flow over the summer and autumn period, and this
in itself may increase the useful yield of an aquifer over
several seasons. These changes are shown for an idealised

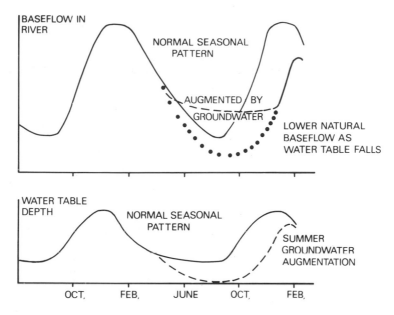

Figure 2. Normal seasonal trends of river baseflow and groundwater depth and the effect on these of augmenting baseflow in summer by groundwater pumping.

system in Figure 2. Actual data for groundwater changes in a very intensively pumped experimental catchment at Thetford, Norfolk during the Great Ouse Groundwater Pilot scheme from 1969-71 are shown in Figure 3. In this study the large additional storage created during 1970 was fully made up during the rather dry winter that followed.

It would be possible, but not in any realistic sense economical, to manage whole aquifer systems in such a way that all base flow was stopped by lowering water tables permanently below the levels at which they would normally discharge and maintaining a pumped base flow of any particular pattern. This extreme form of management would have a considerable ecological impact since many springs, ponds and wet areas would be much drier than normal for substantial periods, and seasonal streams would be much more temporary unless they were artificially supported. The discharge of deep groundwater to form the substantial flow of a river may also change such characteristics as river water temperatures, these usually being made lower in summer and higher in winter, as well as altering dissolved gas contents, mineral content etc.

Current schemes for conjunctive use of groundwater on substantial areas are however usually based on the intermittent use of groundwater in amounts varying with demand but with the total use well within the normal seasonal recharge yields, and with well-sites and discharge points located to diminish ecological effects on riparian land or in the river itself.

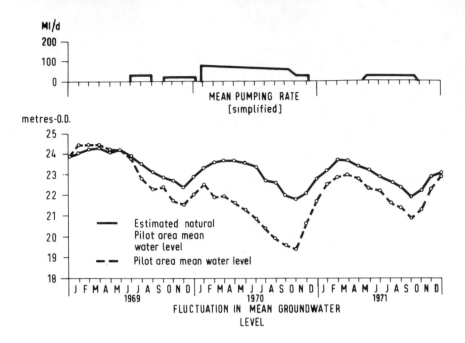

Figure 3. Groundwater level changes in Thetford Pilot
    Groundwater Scheme during an experimental pumping
    programme in comparison with the natural groundwater
    levels estimated from correlation with a nearby control
    catchment.   (after Great Ouse Groundwater Pilot Scheme,
    Final Report, May 1972).

AGRICULTURAL CONSEQUENCES

The agricultural consequences of these schemes tend to fall
into two categories, those concerned with effects on existing
farm water sources and those related to possible drainage
effects on farmland and the crops grown on them.
    Existing water sources may include stock watering or
irrigation points on small streams in addition to farm wells.
These existing sources are protected by law and Water
Authorities, like any other developer, are required to make
good at their cost any detriment to a supply or to provide
an alternative source such as connection to a mains supply
if they cause derogation of supply.   The test pumping requirements linked to applications for licensing new borehole
sources should identify all locations around a new source
requiring alteration, such as deepening wells, lowering
pumps, uprating pump sizes etc. to maintain existing supplies
before any substantial use of a new source is possible.   It
is also common practice in new groundwater development areas
to give an assurance that future developments for local use
in reasonable quantities will not be prejudiced by committing
all available resources for use outside the area.
    The lowering of a water table by groundwater pumping can

result in the drainage of farmland with a naturally high water table providing this is in direct continuity with the aquifer being pumped. In many parts of the world, Holland, Pakistan and Iraq for example, such a means of artificially draining low-lying farmland by pumping groundwater and discharging it elsewhere is part of the normal land management system and in Britain parts of Fenland are artificially drained in most conditions. If however such wetland continues to produce grass during dry summers when other fields suffer drought effects then the loss of this grass would be of some value to a dairy farmer. In practice, however, many farmers given a reasonable certainty of improved drainage of such wet areas would prefer to see such land in regular cultivation or seeded to better quality grass. The same drainage effects on marshland, bogs and wet meadows could result over a period of years in the replacement of plants and animals capable of tolerating such 'wet habitats' by perhaps less interesting forms more suited to the drier habitat. Where groundwater is being mainly used to augment water supplies at times of low natural flow, i.e. in drought conditions, the drainage of these 'wet' sites may not be increased in frequency since it would probably occur naturally in such conditions. However, the drainage effect would be more severe than in the absence of nearby groundwater pumping and the recovery to 'wet' conditions might take a little longer to achieve. Since much of this 'wet' land tends to be in riparian areas, the use of rivers to carry groundwater may itself buffer local water tables against drastic changes and the siting of boreholes such that the major part of their cones of depression are at some distance from stream sides will not only help to avoid large water table changes in riparian areas but

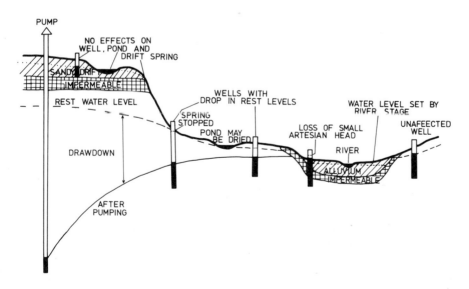

Figure 4. The possible effects of groundwater pumping on neighbouring sites of hydrological interest.

may also reduce recirculation of river water back to the borehole (Figure 4).

Where such drainage effects occur in agricultural soils, it is often forgotten that the amount of water extracted by drainage is quite a small proportion of the water present in a saturated soil. After drainage has ceased the soil is still substantially wet and can yield large amounts of water to plant root systems before the soil becomes physiologically dry to plants, (Marshall, 1959).

Since few plants will root into waterlogged soil (Leyton and Rousseau, 1958; Kramer, 1969), drainage may in some circumstances give plants access to more soil water by deeper rooting than in undrained soils. In addition, drainage may give better soil conditions for cultivation, mineral uptake, weed control and harvesting crops. Drainage can increase yields of crops but the depth of water table which is critical varies with soil type and rainfall amounts (Figure 5).

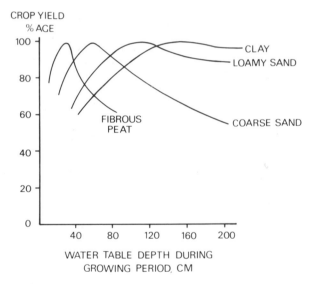

Figure 5. The effect of mean water table depth during the growing season on crop yield in 4 soil types (after Wesseling, 1958).

There is usually no discernible change in the quantity of water held in soils by capillary forces in drained soil layers with water tables already at a moderate depth, say 1 to 1.5 m, if such water tables are lowered by a further 0.5 to 1 m. Laboratory measurements on soil columns on a suction plate apparatus taking many days to reach equilibrium suggest that there should be small effects in some soil types but the cyclical changes of wetting and drying related to rainfall and dry weather patterns mask any small equilibrium changes associated with drainage effects in most field situations.

The upward rise of capillary water from saturated or wet layers to drier layers above is one of the attributes of a shallow groundwater level often regarded as a benefit to the farmer as a means of avoiding the worst effects of summer drought on his crops.   There is often a considerable mis-conception about the role of capillary rise as a contribution to plant water supplies.   Most crop roots are located in the upper 1 m of a soil (Wesseling, 1958;   Wellbank et al, 1974) and there are few roots even in deep rooted crops below about 1.5 m except in some special well-drained circumstances e.g. limestone fissures, deep coarse-grained sands.   To be of value in maintaining crop growth during drought periods, capillary rise needs to be able to provide each day an amount similar to that lost in a normal summer day's evaporation. Where these amounts do not match, the inevitable consequence is a net drying of the root zone and the even more inevitable consequences of this is that rates of unsaturated water flow across even moderately dry soil layers become very slow indeed.   In most soil materials this usually means that water tables must be located within 0.25 to 1 m of the bottom of the main root zone to make any effective capillary rise (Figure 6).   As a general rule water tables deeper than 2 m below surface can be dismissed as being of no consequence to crops.   With trees in this country rooting mainly to about 2.5 to 3 m, water tables more than 4 or 5 m below surface are also of little direct benefit.   In irrigation calculations capillary rise is usually commuted to a constant figure of about 25 mm in a season which, added to the available water in the root zone, gives the basis for the root constant and limiting deficits used by Penman (1963b) and the Meteorological Office in estimating actual evaporation

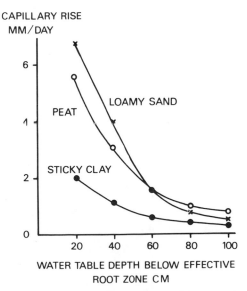

Figure 6.   Relationship between capillary rise into a
    moderately dry root zone and water table depth below the
    root zone in 3 soil types (after Rijtema, 1968).

rates (Anon., 1954a). The sad fact is that our crops and surface vegetation have to harmonize their growth with the relative water changes resulting from current rainfall and evaporation that occur in a layer of about 1 to 1.5 m of soil. Soils which might have water tables high enough to be beneficial by supporting capillary rise at a sufficient rate in mid-summer are likely to be too wet to cultivate in early spring and at risk for harvesting in wet autumns. Farmers recognize this and either do not crop such land or spend large sums draining it. Given care in choice of well sites and suitable monitoring of potentially sensitive areas there need be no serious or long lasting harmful agricultural or ecological changes resulting from sensible programmes of groundwater development.

5

THE EFFECTS OF GROUNDWATER DEVELOPMENT: THE CASE OF THE SOUTHERN LINCOLNSHIRE LIMESTONE AQUIFER

D.B. Burgess and E.J. Smith

Welland and Nene River Division, Anglian Water Authority, Oundle

ABSTRACT

The hydrological characteristics of the Southern Lincoln-shire Limestone groundwater unit are described together with the history of groundwater abstraction. The con-sequences of further development are realised by considering the whole water resource system. This has been guided by the use of field investigations and digital and analogue model simulations. Several water resource management schemes have been considered and a period of temporary overdevelopment is described. The problems associated with the development of a fast response time aquifer are out-lined with particular reference to contiguous river courses and saline water intrusion.

INTRODUCTION

In recent years groundwater sources have become more important to the development of regional water resource systems. The rationale and advantages of groundwater development within the United Kingdom have been presented by Ineson (1970) and Downing et al. (1974). Many of the investigations into the management of such developments have been the responsibility of the River Authorities and Water Resources Board and more recently, the Regional Water Authorities under the terms of the Water Resources Act of 1963 and the Water Act of 1973. As a result, many ground-water investigations are now carried out at the regional scale within the context of the complete hydrological cycle. Such changes in the scale of investigation have been parallelled by the extension and increased precision of hydrometric measurements, the improvement of geophysical and geochemical techniques and an increase in the sophis-tication of analogue and digital models employed in the regional analysis of groundwater systems.

The optimal development of groundwater resources is generally associated with abstractions from the aquifer balancing the recharge to it (Ineson, 1966). When abstractions exceed recharge, a number of undesirable effects may result. These range from the regional fall in rest water levels, as in the chalk of the London Basin (Water Resources Board, 1972) to the intrusion of saline water, as into the Permo-Triassic sandstones of south Lancashire (Bow et al., 1969). Although the effects of long-term over-development are undesirable, it has recently been acknowledged that a policy of controlled dewatering may be acceptable over a limited period of time until other sources are introduced (Downing et al., 1974). This paper outlines such a short period of overdevelopment of the Southern Lincolnshire Limestone groundwater unit and the groundwater dependent river system.

The salient features of the groundwater resources of the Lincolnshire Limestone have been outlined by Downing and Williams (1969) for the period 1960-67. In the late sixties and early seventies abstractions have significantly exceeded the recharge of the aquifer. The consequences of such over-development have been the subject of a series of investigations (Anon., 1971, 1972, 1973, 1976).

HYDROGEOLOGICAL CHARACTERISTICS

The western boundary of the Southern Lincolnshire Limestone groundwater unit is a groundwater divide coincident with the topographic divide between the Witham and Glen catchments (Figure 1). The southern limit is along the line of the Marholm-Tinwell Fault which acts as an impermeable boundary (Figure 2). Between the two boundaries, in the Stamford to Witham area, the aquifer is continuous and no obvious natural divide exists. However, recent exploratory work has indicated that groundwater movement is not significant in this area. The northern boundary trends roughly west-east and is associated with a recharge dome located to the east of Grantham. In general this appears to be associated with disturbances of the Syston-Dembebly monocline. The eastern boundary, for water resource purposes, is dictated by the occurrence of saline, connate water. The actual boundary is the limit of potable water and is taken as the 250 ppm isochlor.

The Lincolnshire Limestone is a hard compact limestone which exhibits considerable lateral and vertical facies variation. It forms part of the Inferior Oolite Series of the Jurassic which outcrop from Humberside in the north to Northamptonshire in the south. The area of interest extends south from Grantham to Stamford and eastward into the Fenland of Eastern England. The outcrop of the Southern Lincolnshire Limestone unit is some 12 to 14 km wide. It dips gently eastwards at approximately 1^{O} and is traversed by west to east trending folds and associated faults. The Limestone is 40 m thick near Grantham and thins south-eastwards to 15 m to the east of Spalding. Although lateral and vertical variations in lithology are common, the pattern

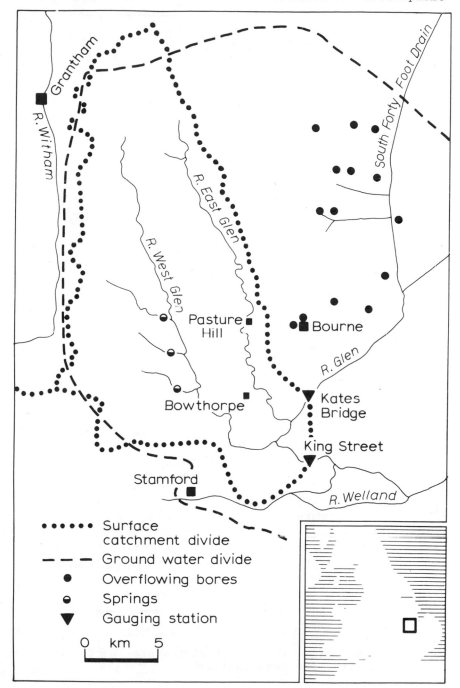

Figure 1. The location of the Southern Lincolnshire
Limestone aquifer and hydrological features of the area.

Legend:
•••••• Surface catchment divide
— — — Ground water divide
● Overflowing bores
◒ Springs
▼ Gauging station

0 km 5

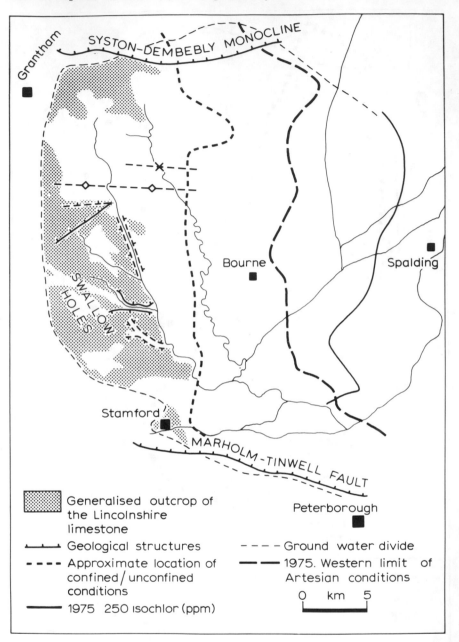

Figure 2. The main hydrogeological features of the Southern
Lincolnshire Limestone aquifer.

of groundwater movement and storage is dominated by a well-
developed system of fissures particularly in the confined
area. Intergranular permeabilities are insignificant
regionally (Anon., 1973). However, in the outcrop, where

poorly cemented oolitic limestones occur, intergranular
flow may be locally important.

In general the Limestone overlies relatively impermeable clays and silts of the Lower Estuarine Series.
However, these confining strata are not persistent
regionally and the limestone is often in hydraulic continuity with the Northampton Sandstone. The sandstone aquifer
does not contribute significantly to the overall resources

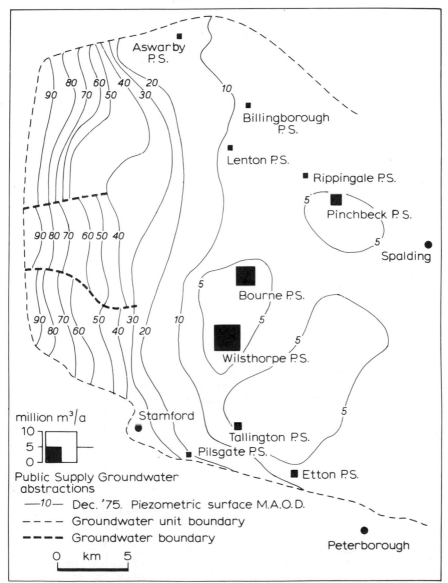

Figure 3. The December 1975 configuration of the groundwater table and the principle groundwater abstraction
centres.

of the groundwater system although it may have some bearing on the overall quality. Under these conditions Lias clay acts as the basal confining strata.

The Upper Estuarine Series of clays and shales overlie the Lincolnshire Limestone in the east and approximately 5 km downdip of the junction confined conditions of groundwater flow predominate. Within the Fenland area, artesian over-flowing conditions are common.

The regional hydraulic gradient is from west to east in the outcrop although the water table surface is complicated by a series of west to east trending faults and two deep glacial buried channels (Wyatt et al., 1971). In addition valley bulging along the River West Glen and its tributaries has disrupted the regional flow pattern. Consequently, partially isolated aquifer blocks exist and strong spring flows emerge. The water table gradient ranges from 1:40 to 1:350 and from north-west to south-east, to south-west to north-east (Figure 3). With the onset of confined conditions there is a significant reduction in gradient and within the confined zone gradients range from 1:500 to 1:7000. Although the natural direction of groundwater movement is predominantly from west to east there is significant interference from the principal centres of groundwater abstraction. A complex piezometric surface exists with assymetrical cones of depression complexly integrated with remnants of the natural groundwater surface.

The annual range of the water table varies considerably throughout the outcrop from less than 1 m to in excess of 4 m. There does not appear to be any pattern to the spatial variation in the range, although it is obviously a function of *inter alia,* the size and density of fissures within the borehole profile and their lateral regional interconnection. The range of fluctuation is influenced by abstractions downdip in the confined zone although the interference effect at any particular borehole cannot be determined precisely at present.

The fissured nature of the Limestone outcrop and the presence of numerous hydraulic boundaries and discontinuities precludes the determination of aquifer characteristics by conventional pump test analyses. Analysis of natural hydrograph responses (Headworth, 1972) and regional analogue/digital model analyses (Fox and Rushton, 1976) indicate a regional transmissivity value of less than 500 m^2/day which is superimposed by discrete high values of the order of 1000 to 2000 m^2/day. The average regional specific yield is of the order of 0.05. Where confined conditions prevail transmissivity ranges from 2500 m^2/day in the north to 10000 m^2/day locally in the south (Anon., 1972, 1973). The corresponding storage coefficients range from 5×10^{-3} to 5×10^{-5}.

The area of outcrop is partially covered by boulder clay and Jurassic clays which restrict the direct infiltration of precipitation. However, runoff from these deposits enters the Limestone through several nests of swallets associated with the boundary of the boulder clay (Hindley, 1965). Groundwater discharge in the outcrop occurs at a number of springs emerging from blocks of limestone isolated as a result of valley bulging. Groundwater

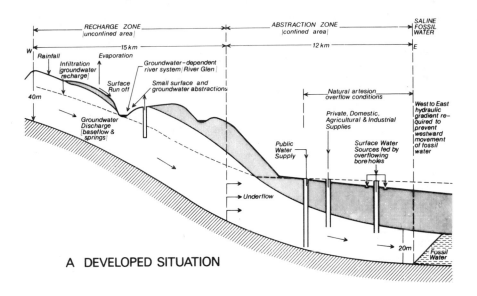

A DEVELOPED SITUATION

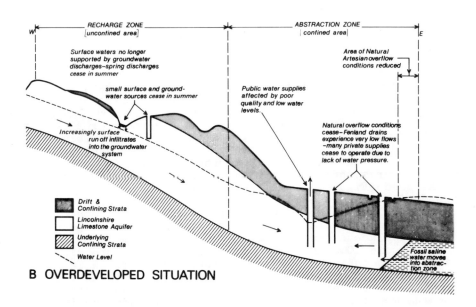

B OVERDEVELOPED SITUATION

Figure 4. Schematic diagrams of the hydrological cycle of the Southern Lincolnshire Limestone aquifer showing: (A) the developed situation, (B) the overdeveloped situation.

that does not contribute to the baseflow of the river system flows downdip and into storage in the confined zone (Figure 4 A).

Within the outcrop and for approximately 15 km into the confined zone the groundwater is hard as a result of the dissolution of calcium carbonate and the quality is adversely affected in some places by runoff from a predominantly agricultural basin. Further downdip, the water is softer and the dominant cation is sodium. Carbonate is replaced by chloride as the dominant anion towards the east (Lawrence et al., 1976; Downing and Williams, 1969). In the extreme eastern confined zone the groundwater becomes increasingly saline with sodium chloride predominating. The main abstraction area is therefore restricted to the western and central portions of the confined zone. This is suitably remote from the saline water to the east and from direct contamination from the recharge waters updip. A substantial volume of the good quality, 'ancient' water exists within the areas of abstraction and this has the effect of diluting poorer quality water drawn in during periods of high abstraction (Downing et al., 1977).

THE HISTORY OF GROUNDWATER ABSTRACTION

The occurrence of artesian overflowing conditions and good quality water within the central confined zone has resulted in a long history of abstraction. In 1880 the piezometric surface was recorded as being 15-20 m above ground level (Woodward, 1904). These early records also indicate the discrete fissured nature of the Limestone; in some areas 'dry' boreholes are mentioned while in others water bearing horizons capable of yielding 5 to 10 Ml/day by natural overflow were found (Isler & Co., 1893).

Since the 1920s, abstraction has been dominated by public water supply undertakings with major source works within the confined area. Minor source works were located on the outcrop but proved to be less reliable in terms of yield and quality. By the 1960s public water supply abstractions were increasing at the rate of 4.7% per annum with a corresponding reduction in the number of source works used (Figure 3). In 1970 eight source works within the confined zone accounted for 86% of the total licensed abstraction from the aquifer (Figure 5A). The remaining abstractions were for industrial and agricultural purposes and took place at approximately 300 sites spread over an area of some 1000 km. In addition, several boreholes within the confined zone were allowed to overflow in order to support flows within the Fenland drains. Total flows from these were estimated at 13 Ml/day. These contributed towards meeting licensed surface water demands and in preventing the inland incursion of saline water past the tidal limit.

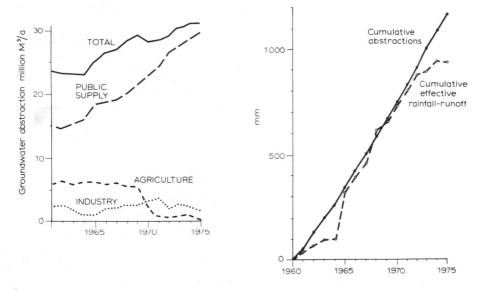

Figure 5. The water balance, 1961-75, of the Southern
      Lincolnshire Limestone aquifer : (A) groundwater
      abstractions, (B) cumulative water balance.

CHARACTERISTICS OF DEVELOPMENT

In order to assess the stage of development of the
Southern Lincolnshire Limestone, an accurate evaluation of
its safe yield is required. Conventionally, the yield of
an aquifer is taken to be equivalent to its long term
average natural recharge, with the assumption that the
quantity of water stored in the aquifer is large enough to
balance out any variations in underflow and abstraction
(Twort et al., 1974).
      The recharge area is drained by the River Glen which
has been gauged at Kate's Bridge continuously since 1961.
The area of this catchment approximates to the total
recharge area (Downing and Williams, 1969). Consequently,
the amount of underflow to the aquifer may be assessed from
a water balance (Figure 5B), the method of which is
described in Ineson and Downing (1964). For the period
1961 to 1975, the water balance is:

Underflow + Total Runoff Rainfall - Actual Evapotranspiration
$$\text{64 mm} \qquad \text{96 mm} \quad = \quad \text{612 mm} \qquad \text{452 mm}$$

Two problems exist in using this particular form of resource
assessment. First, the effective rainfall of the right hand
side of the equation is subject to various forms of error.
Estimates of areal rainfall over a catchment are subject to
errors which arise from point measurements (Rodda, 1967)
and from techniques used to compute the area estimate from
the various points (Hutchinson, 1970). The limitations of

estimating actual evapotranspiration for recharge estimates
are also well documented (Headworth, 1970). The second
problem lies with the interaction of the total runoff and
underflow terms. Prior to significant abstractions from
the confined area, it is reasonable to assume that the
majority of the infiltration over the recharge area
reappeared as springs and baseflows to the River Glen. No
natural outlet exists to the east, although leakage to
contiguous strata has been assessed at 5 Ml/day (Downing
and Williams, 1969). The subsequent abstractions from the
confined zone have led to a regional lowering of the
piezometric surface which in turn has resulted in the progressive capture of the baseflow component of the River Glen,
as well as the occurrence of seasonal influent conditions
along its course. Thus, while both sides of the equation
are balanced the left hand side represents a progressive
increase in the amount of underflow at the expense of total
runoff.

The baseflow of the River Glen is important in maintaining summer flows in the Fenland area and therefore
successful management of the aquifer is based on producing
an acceptable balance between surface and groundwater
sources. Over-development of the aquifer can only be
assessed in terms of the total impact on the water resource
system.

Some indication as to the various states of balance
between the amount of water abstracted and the magnitude of
the underflow component can be deduced from records of the
piezometric surface (Figure 6). Both maximum and minimum
annual levels have a downward trend, but with minimum levels
declining at a faster rate from the 1920s to the early
1960s. This is indicative of an imbalance between underflow and abstraction and can be considered as a period
during which a dynamic balance was being attained. This was
achieved by a gradual updip extension of cones of influence
and a progressive capture of runoff. By the 1960s an
adequate proportion of runoff had been induced to underflow
and a suitably inclined west to east hydraulic gradient had
been established between recharge and abstraction. During
this period underflow was sufficient to balance abstraction
without significant interference to flows in the River Glen.

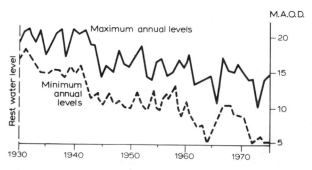

Figure 6. The range of annual fluctuations of the
piezometric level at Bourne Pumping Station, 1930-1975.

CHARACTERISTICS OF OVERDEVELOPMENT

While it may be argued that the 1960s represented a period
of balance between abstraction and recharge, the combination of a temporary increase in demand together with below
average rainfall for the years 1970-75 resulted in a period
of overdevelopment. A progressive lowering of the
piezometric surface ensued and this was exacerbated by a
rapid updip expansion of cones of depression from around
the principal sources of abstraction. The principal
consequences of this were a contraction of artesian overflowing conditions eastwards, a westward updip migration of
the zone of saline water and a reduction of flow in the
River Glen (Figure 4B).

Prior to this period, artesian conditions were general
over the Fenland area with only local contractions around
the major abstraction boreholes. During the period 1970-75
the piezometric surface fell below ground level over the
western confined area during the summer months. This
resulted in the cessation of flow of many Fenland springs
and unregulated boreholes. Summer flows in Fenland drains
were subsequently reduced and these were insufficient to
prevent the incursion of saline water along reaches of the
River Glen. An increasing shortage of water for spray
irrigation purposes was experienced. In addition many of
the small domestic groundwater abstractions suffered
reductions in natural artesian pressure which in the past
had been adequate for distribution purposes. The rate of
decline of the piezometric surface was aggravated further
by a significant seasonal lowering of the water table updip
of the abstraction zone. A significant reduction of the
saturated thickness subsequently inhibited replenishment of
the abstraction zone during periods of groundwater level
recession.

The natural form of the piezometric surface is to fall
gently eastward and this has inhibited updip movement of
the saline water into the abstraction zone prior to 1960.
However, subsequent increased abstraction has reversed the
hydraulic gradient in the vicinity of the fresh/saline
interface during periods of groundwater level recession.
A net westward movement of the 250 ppm isochlor of up to
2 km had been detected in places in 1975. Several small
domestic sources have suffered from water quality problems
and one small public water supply source has had to be
abandoned (Anon., 1976).

The progressive capture of the baseflow component of
the River Glen has already been referred to in connection
with the overall surface/groundwater balance. During the
1960s groundwater discharge to the River Glen was sufficient
to maintain summer flows at an 'acceptable' level of discharge. The subsequent period of overdevelopment resulted
in a progressive lowering of the water table between the
abstraction zone and the River Glen and a spatial and temporal increase in the occurrence of influent conditions. A
progressive reduction of the annual yield of the watershed
has ensued and since 1970 a 80% reduction in the naturalised
dry weather flow has occurred (Figure 7A). The extent of
this effect can be depicted by a parallel shift in the

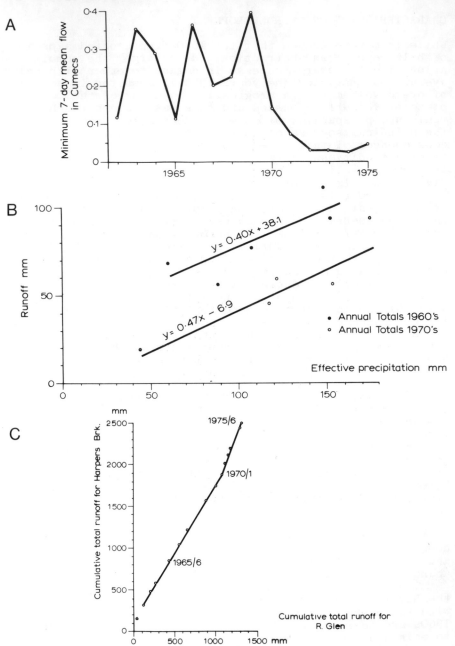

Figure 7.   The annual variations in the yield of the Glen
            catchment 1961-75.
            (A)   Trend in the minimum seven day mean flows.
            (B)   Changes in the total runoff vs. effective
                  precipitation relationship.
            (C)   Double mass analysis of annual total runoff:
            Harpers Brook at Old Mill Bridge vs the River Glen at
            Kate's Bridge and King Street.

effective rainfall/runoff relationship for two periods of
similar annual rainfall populations in the early 1960s
and 1970s (Figure 7B). This change in the relationship is
significant at 0.10 as tested by an analysis of covariance
(Riggs, 1968). A double-mass analysis of the runoff of
the River Glen with that of a nearby undeveloped impermeable catchment, Harpers Brook (NGR: SP 983719) indicated a consistent relationship throughout the 1960s with a
break occurring from 1970 onwards (Figure 7C).

The capture of surface runoff from the Glen catchment
which is predominantly rural in character and intensively
cultivated, has resulted in a deterioration in the quality
of the recharge water. In particular there is recent
evidence of seasonal increases in the concentration of
chloride, sulphate and nitrate in the water at outcrop.

RESULTS FROM GROUNDWATER MODELS

The problems associated with the assessment of the safe
yield of a fissured aquifer with boundary conditions in
close proximity, together with the obvious delicate balance
between surface and groundwaters, prompted the use of
digital and analogue groundwater simulation models to
evaluate present and proposed methods of groundwater
development. Initially, difficulties were incurred in
reproducing the regional fluctuation of the piezometric
surface in response to variations in recharge and abstraction. However, as more data from geophysical investigations and pump test analyses revealed the nature and
disposition of fissuring within the limestone, the aquifer
characteristic distribution and the recharge were modified.
The resulting models were able to simulate historical
events with increasing degrees of accuracy.

A posteri models of the present scheme of aquifer
development were produced and a reasonable agreement with
observations for the period 1961 to 1975 were obtained.
Calibration procedures indicate that discrete horizons of
high transmissivity are more extensive within the Glen
catchment than previously proposed. These horizons are
associated with a well developed fissure system which
facilitates rapid recharge (Fox and Rushton, 1976). From
interpretation of the response of the aquifer updip of the
abstraction zone it is apparent that the unconfined/
confined 'boundary' is not coincident with the intersection
of the water table with the top of the Limestone. Indeed
the top sequence of Limestone, which can be of the order
of 2 to 4 m in thickness, confines flow at deeper horizons.
The occurrence of confined conditions of flow in fissures
further updip than previously anticipated result in a more
widespread interference from abstraction than previously
envisaged. In addition, it is possible that the bulk
storage capacity of the aquifer may have been overestimated in previous studies. The combination of high
transmissivities and small storage coefficients, particularly in the main abstraction zone, results in a fast
response system. As a result boundary conditions are
significant throughout all stages of development and in

particular when overdevelopment proceeds.

Simulation models have subsequently been employed as technical aids to water resource management decisions. The effects of various forms of groundwater development have been reviewed and several significant results have been forthcoming. Earlier investigations proposed the development of groundwater storage within the confined zone as a necessary prerequisite to artificial recharge from nearby inland surface waters, (Downing and Williams, 1969). Model results from this scheme suggest that the fast response time of the aquifer would preclude isolated aquifer development of this nature. Drawdown effects extend rapidly updip to interfere with surface sources and acute reversals of hydraulic gradient in the east encourage the encroachment of saline water into the abstraction zone. The sensitivity of the system consequently encouraged a more conservative approach and a number of spatial and temporal patterns of abstraction were considered. Overall the results did not indicate any advantages in modifying the present pattern of development.

The models were further used to predict the consequences of a continuation of the historical rate of growth in abstraction of approximately 5% per annum. Several significant features emerged. In the confined aquifer a progressive reduction of the piezometric surface would result in an entire loss of artesian overflowing conditions and encroachment of saline water into the abstraction zone. The latter effect would be more acute if flow occurred differentially along discrete fissures. Of the abstracted water a substantial proportion would be derived from the Glen catchment at the expense of unconfined storage and river flow. If smaller storage capacities are incorporated in the recharge zone an accelerated updip migration of the cones of depletion would occur. Consequently, influent river conditions would prevail for longer periods and the runoff would be captured entirely during periods of below average rainfall. During periods of water table recession significant reductions of the saturated thickness would occur updip of the abstraction zone. Dewatering of major fissures would reduce the regional transmissivity thus making groundwater storage in the recharge zone increasingly unavailable to the abstraction zone. This would lead eventually to an accelerated rate of decline of the piezometric surface.

CONCLUSIONS

The implications for water management of these investigations are twofold. First, the traditional methods of calculating the safe yield of an aquifer from the long term natural recharge are best applied prior to significant groundwater abstraction. Once abstraction has been initiated, the successful development of the groundwater system depends on an accurate assessment of the amount of underflow that may be induced from the recharge area without seasonally dewatering it. An understanding of the dynamic and interactive responses of the groundwater and the contiguous river system is required; this can be achieved through the employment of conceptual deterministic models.   Second, the principal effects of groundwater development within the Southern Lincolnshire Limestone have been parallelled by a more profound appreciation of the hydrological variability of the aquifer and the development of sophisticated analytical techniques.   In particular, a fuller appreciation of the hydrogeological implication of a fissured system has led to a better under-standing of the whole groundwater cycle within the aquifer. As in other studies, the incorporation of these findings into simulation models has provided insight into aquifer development as well as aiding future management decisions (Birtles and Nutbrown, 1976).

It has been suggested that artificial recharge of the Lincolnshire Limestone in conjunction with river regul-ation from the recently completed Rutland Water reservoir, would provide the most efficient method of exploiting the groundwater system (Jamieson et al., 1974).   However, as in other aquifer developments, the practical feasibility of such a scheme rests with an adequate understanding of the natural recharge mechanism.   Present investigations are directed towards tracing the subterranean path of water entering the Limestone while a hydrochemical investigation of the indigenous groundwater will provide a guide as to the required chemical and biological quality of the recharge water (Lawrence et al., 1976, Downing et al., 1977).

ACKNOWLEDGEMENTS

This paper is published with the permission of the Anglian Water Authority.   The opinions of this paper are those of the authors and not necessarily those of the Authority.

6

THE HYDROLOGICAL IMPACT OF AFFORESTATION IN GREAT BRITAIN

W.O. Binns

Forestry Commission, Research and Development Division, Farnham, Surrey

ABSTRACT

Forestry, which occupies 8.5% of Britain's land area, reduces water yield because it reduces albedo, increases air turbulence in the canopy, traps precipitation by interception and draws moisture from depth via deep roots. The hydrological effects of different tree species and the four phases of forest operations are described in terms of runoff rate, runoff volume and water quality. Measures to minimise the reduction in water yield can cost the forester very little. It is concluded that the effect of forestry on water yields will be determined more by policy on land acquisition and use than by the actual forestry operations themselves.

FOREST AREAS

The first Census of Woodlands in Great Britain to follow the creation of the Forestry Commission in 1919 estimated that there were 573 thousand hectares of productive high forest in Britain. Table 1 shows that this was about 2.5 per cent of the total land area (Forestry Commission 1928). By 1976 the area of managed forest land (not all planted) had risen to about 1.69 million hectares, of which 915 thousand hectares were Forestry Commission land and 576 thousand dedicated or approved land (Forestry Commission, 1976). In addition there are about 300 thousand hectares of woodland classed as unproductive, bringing the total forest and woodland area to about 2 million hectares or just over 8.5 per cent of the land area. From 1970 to 1976 the Forestry Commission acquired on average just over 13 thousand hectares of land a year, mostly in Scotland.

Man's impact on the hydrological cycle

Table 1.   Census of Woodlands 1924.

		area 000 ha	% of total land area
Potentially economic	(High Forest	573	2.5
	(Coppice and coppice - with standards	214	0.95
	(Scrub, felled and devastated	327	1.45
	uneconomic	83	0.35
	Total woodland	1,197	5.25

Figure 1 shows that although there is, in total, a substantial area of managed forest in the lowlands of England, the bulk of the large forests lie in the uplands. There has been relatively little new planting in the lowlands, many of the larger forests in southern England being old Royal forests, though a few have been planted on impoverished sandy or chalky soils which were considered useless for agriculture in the inter-war years. It follows therefore that the hydrological changes brought about by afforestation will have their greatest impact in the uplands, which are also important sources of surface water for much of Britain (Figure 2).

THE NEW UPLAND FORESTS

The land in the uplands which has been planted since 1919 has been in the main rough grazing, grouse moor and unused peat bogs. Only small areas of old broadleaved woodland, mainly oak and birch, have been replaced with conifers. Great changes have resulted from all these plantings. A much more vigorous vegetation type has replaced an impoverished one, often with the assistance of mineral fertilisers which eliminate the marked deficiencies of important plant nutrients brought about largely by a purely extractive land use, frequently including muirburn. In addition, road making, draining, and cultivation have been widely carried out. These two aspects of afforestation, the substitution of a very different vegetation type and the hydrologically important operations associated with managed forestry, are discussed in this paper.

EVIDENCE FOR THE EFFECT OF FORESTS ON HYDROLOGY

Studies in various parts of the world, most importantly in the United States and South Africa, show convincingly that the removal of forest increases the stream flow in well

watered catchments and that allowing it to grow again reduces
the yield to the original level; the data up to 1966 were
reviewed by Hibbert (1967). Britain had been little involved
in controversies over the effect of forest on water yield
until Law asserted to the British Association in 1956 that
his forest lysimeter of Sitka spruce in Lancashire had
evaporated 275 mm more water in a year than the surrounding
grassland (Law, 1957a). Several criticisms were immediately
levelled at this work. First, it was felt that the
evaporation would be artificially high because of the small
size of the stand, in other words there would be a large oasis
effect. Second, since evaporation is a physical process, it
seemed that the total evaporation ought to be the same for
all vegetation types. Third, although it was accepted that
water was intercepted by the canopy of the trees and later
re-evaporated without ever reaching the ground, it was argued
that the energy used in this process would not then be
available for transpiration. On reflection it seems odd that
there should ever have been much doubt about a greater water
use by forest, since it must have been regularly observed by
every humble woodman who worked on heavy soils, even in
relatively low rainfall areas, that the soil became much
wetter after the forest was felled, but that the original
balance was restored once a new crop of trees had grown up,
since that is what we see today. Yet Brown (1882) and
Nisbet (1905) are quite categorical in their assertions that
forests increase the soil moisture and water supply. Laurie
(1956), in theoretical discussions of the cost of water loss
associated with forestry, was considering a figure as low as
40 mm a year.

The Institute of Hydrology (Clarke and Newson, in press;
Clarke and McCulloch, Chapter 7) has recently summarised its
observations over 5 years at the Plynlimon catchments on the
headwaters of the Wye and Severn. They show that from 1970-
75 the Wye catchment, under grass, lost 18 per cent of the
mean annual precipitation of 2415 mm, while the Severn catchment, under forest, lost 30 per cent of 2388 mm. The
difference between the two is about 280 mm - remarkably close
to Law's figure. They conclude that the effects of the 80%
cover of trees on the 8.7 km^2 Severn catchment is to use an
additional 2.2 million m^3 of water a year. This is a
considerable figure and should be taken into account when
planning land use in the uplands.

MECHANISMS OF EVAPORATION

Forest uses more water than pasture or other low vegetation
because of four factors. First, forest has a lower reflectivity
(albedo) so it absorbs more incoming radiation. Pereira (1973b)
quotes figures from Ångström of 26 per cent for grass, 17.5
per cent for oak wood and 14 per cent for pine forest. Stewart
(1971) recorded 9.1 per cent for pine forest at Thetford
(Norfolk and Suffolk) and at the Coalburn catchment in
Northumberland the pasture measured 15 per cent before ploughing
and 13 per cent afterwards (Institute of Hydrology, 1973). This
lower albedo means there is more energy available to evaporate
water from the forest. Second, the forest canopy is rough
and deep which causes a greater mixing of the air passing

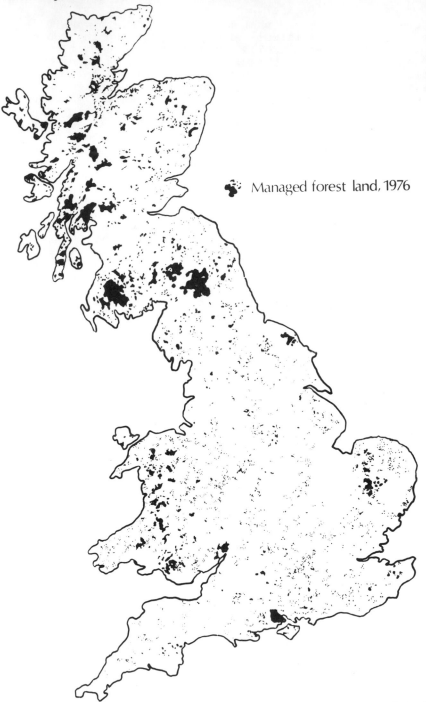

Managed forest land, 1976

Figure 1. Managed forest land in Britain. Foresty Commission
land as at 31.3.76, private forest land as at 30.9.75.
(Note: only those islands with appreciable areas of managed
forest have been included; some of the smaller forest areas
do not show up on this scale). Based upon an Ordnance Survey
map with the permission of the Controller of Her Majesty's
Stationery Office. Crown Copyright Reserved.

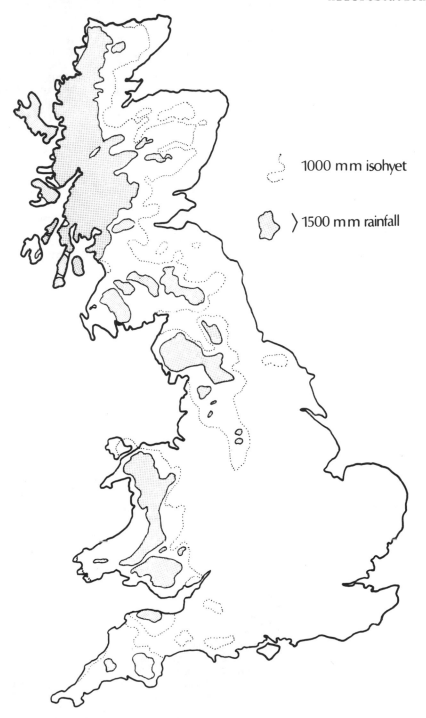

1000 mm isohyet

1500 mm rainfall

Figure 2. Land in Britain receiving more than 1000 mm and 1500 mm precipitation.

through it. This results in more efficient transfer of water
vapour and thus greater evaporation. Third, an appreciable
proportion of any one storm, commonly up to 2 mm (Kittredge,
1948), is held on the leaves or needles of complete forests.
This evaporates after rain ceases, the energy being advected
in the incoming air stream which is cooled in the process.
Some very high interception values have been recorded, for
example Delfs (1967) found a maximum value in Norway spruce
of 17.5 mm for a single storm. Finally, some forest trees
are deeper rooting than other forms of vegetation and in
long-continued drought they will be able to transpire at the
normal rate after shallow-rooted plants have been forced to
reduce their transpiration rate considerably. This will not
greatly affect either surface runoff or ground water supplies
at the time, but it means there is a greater soil moisture
deficit to be made up when the drought breaks. Ground water
does not therefore start to flow again so soon from the forest
as it does from pasture and, unless the infiltration capacity
is exceeded by heavy storms, rain will leave by sheet runoff
from pasture sooner than it will from forest.

FOG DRIP AND SNOW MELT

It has been claimed that forests increase rainfall and it
does seem that in certain, rather unusual conditions, fog
drip may substantially increase the precipitation within the
forest. The phenomenon seems uncommon in Britain although
lingering doubts may remain amongst those who have been
thoroughly wetted by drips in the forest, yet remained dry
outside it. Russian evidence (e.g. Rakhmanov, 1966) has
often been quoted to suggest that forests increase the runoff.
However, in the region from which this evidence comes, almost
all the precipitation falls as snow and the forest traps and
stores more snow than other land. These conditions do not
occur in the U.K. Penman (1963a) summarising all the evidence,
concludes that although forests may affect where rain and
snow fall, there is no evidence that they alter the total
amount of precipitation in a region.

TREE SPECIES

It has been shown clearly that trees use more water than does
pasture. It has also been suggested that deciduous trees,
because they will intercept less in winter and have a shorter
season of transpiration, will use less than evergreens (e.g.
Delfs, 1967). Stem flow, in contrast, may be very high in
deciduous trees in winter, so methods of measuring interception are very important (Reynolds and Henderson, 1967).
In fact deciduous trees will not grow productively on any
scale in the uplands, although larches may be used on less
exposed sites both for variety and because they give a rapid
early yield of timber. It has always been supposed that
there will not be much difference between different species
of evergreens, but Fourt and Hinson (1970), in an investigation
in south-east England, suggest that Corsican pine dries the
soil more deeply and intensely than Douglas fir and therefore

uses more water during deficit periods. They consider that
the distinctive canopy structure and leaf pattern of these
two species may be responsible for this effect. Although these
two species are not grown much in upland forests, the same
differences of canopy structure and leaf pattern are shown
by Sitka spruce and Lodgepole pine, the two main upland species.
It follows therefore that a pine forest may well use more
water than a spruce forest.

FOREST OPERATIONS

Considered in their effect on hydrology, forest operations
may be divided into four phases. The first is establishment,
which may include the exclusion of stock by fencing, ploughing,
removal of a proportion of the vegetation by herbicides,
building the primary road network, and of course planting.
This phase is considered complete when the crop closes canopy
and shades out other plants. The thicket phase lasts from
closure of canopy until access is made into the crop for the
first thinning. The pole phase normally involves periodic
removal of a part of the crop and imposes a cyclical
modification on the hydrology. This phase usually sees the
completion of the permanent network of roads and extraction
routes. The fourth phase is the clear-felling of the final
crop of trees, followed by replanting and a return to phase one.

Planting new ground

Land newly acquired for planting will first of all be fenced
and normally stock will be removed. The effects of this
enclosure may be quite considerable, particularly if the
pasture has been heavily grazed. The vegetation will become
taller and the water yield is likely to decrease a little,
but this change is small compared with the effect of the ground
preparation which normally precedes planting. Small two-year
transplants are standard in the uplands, though on deep peat
8-12 week-old tubed seedlings may be used (Low, 1975). Small
trees are used because large trees are expensive to grow, time-
consuming to transport and plant, are unstable in windy upland
areas, and often die. The 15-20 cm tall Sitka spruce trans-
plant, which is the most important species in upland forestry,
costs as little as £25 per 1000 (1977 prices) and it survives
very well. A man can plant over 1000 transplants or up to
4000 tubed seedlings a day on favourable ground (Figure 3).
However spruce grows best when there is no competition
from other plants and on almost all upland soils some ploughing
is normal. If the soil is an upland brown earth or ironpan,
it may be ploughed to improve vertical drainage and suppress
weeds with the tree being planted on the top or side of the
weed-free plough ridge. If the alignment of the ridges is
just off the contour then the hydrological effect is slight,
but if up and down the hill there can be considerable increase
in rapid runoff and erosion, particularly in the first few
years following ploughing. Unfortunately the slope of the
ground may be so great as to render ploughing near the contour
hazardous (Figure 4).

Figure 3. Shallow double throw ploughing to provide ridges
     on peaty ground. Good local weed suppression and easy
     planting, but rapid run-off of rain falling into the
     furrow. (Forestry Commission Photo B6399.)

     If the soil is wetter, for example classed as a peaty
gley, surface water gley, or deep peat, then ploughing for
drainage takes the place of the cultivation described above.
Double-throw ploughs have been increasingly used and the
plough ridges give weed-free conditions for some years while
the furrows provide the primary drainage network. The depth
of the primary furrow will vary depending on the need for
drainage. The effect of this drainage can be quite dramatic
on the deeper peats and a great deal of water immediately
leaves the hillside in the new drain channels. There is
usually a secondary system of cross drains to intercept the
water from the plough furrows and this forms the permanent
drainage system of the forest. In gently rolling country the
primary ridges and drains run down the slope and the cross
drains are placed just off the contour. The consequence of
aligning primary cultivation down slope is that the drainage
effect of these furrows is small and limited to a small area
either side of the drain. In contrast the secondary drainage
system intercepts not only the water flowing down the primary
drains but also drains any sub-surface flow from the lands.

Figure 4. Single throw downhill tine ploughing, with a
cross drain just off the contour. Good local weed
suppression, easy planting and improved vertical perco-
lation, but likelihood of increased erosion for some
years. (Forestry Commission Photo B6258.)

Trees are usually planted on the tops of the ridges or in a
step in the side of a ridge on ploughed ground.

Road construction is delayed as long as possible, but the
beginning of the main permanent network is usually established
when planting begins. Road construction can entail cutting
and filling, the bridging and diverting of existing streams,
the blasting of rock outcrops, and the construction of roadside drains with culverts at intervals, the last to prevent
the road being washed away in heavy rain. If the drainage
system has been well designed then the road system should
only alter the drainage locally, but there is likely to be a
considerable effect on water quality during construction.

Herbicides are not usually needed for the first few years
on ploughed ground and the existing vegetation on those parts
of the lands not covered by the plough ridges will continue
to grow normally. An exception is where grasses are the main
weeds and on such sites herbicides may be used in the second
or even the first year.

The hydrological effects of these operations will depend
on whether cultivation or drainage has been used. The

63

consequences of cultivation are small, since infiltration in most of the furrows usually improves while a small number may act as drainage channels if they happen to overlie impervious material. Drains however have the obvious effect of removing water very rapidly and, because up to 20 per cent of the land may consist of ditches, 20 per cent of any storm will fall directly into these ditches and leave the site within a very short time. Thus the hydrograph on a recently drained catchment will show a more rapid response to rainfall then before ploughing. The effect of ploughing has been followed at Coalburn, Northumberland (Institute of Hydrology, 1973) but the full results for the early years have not been published. As the trees grow towards the thicket stage they evaporate more and more water but to begin with this only compensates for the suppression of the natural vegetation. Grass and other plants grow over the ridges and furrows of cultivation and over the ridges from drains, which starts the return to a less flashy hydrograph.

Fertilisation with phosphate is standard in the uplands with the exception of some brown earth and ironpan soils and today is almost always done from the air. If fertilisation follows ploughing a proportion of the fertiliser is bound to fall directly into any drains from which it will be washed into streams and rivers, or reservoirs. If it precedes ploughing, the loss will be smaller, but still important. At Coalburn, where the fertiliser had been broadcast from the air before ploughing, the phosphate concentration in the water leaving the catchment rose to a very high level during the ploughing operation but within four months had fallen to about ten times the original, very low concentration (Institute of Hydrology, 1973). Suspended solids also increased greatly at Coalburn, by three orders during ploughing, but dropped to two orders greater after a few months (Painter et al., 1974). It is expected that the long-term values will fall further but will probably remain somewhat higher than in the original pasture. If the drainage system has been well designed then only minor repairs and alterations should be necessary. These are generally only on a small scale compared with the initial work but will produce further loads of solids to be carried away with the drainage water.

During the establishment period some weeds may begin to compete sufficiently with the trees to reduce their growth rate, in particular the common heather, *Calluna vulgaris*. Herbicides, in particular 2,4-D, are used to kill the heather and restore the normal height growth. This herbicide, while not particularly toxic (Turner, 1977), has an unpleasant taste and very small concentrations can taint drinking water. For this reason it has hardly ever been used from the air (Brown, 1975). Application is increasingly with ultra low volume (ULV) equipment and the chance of polluting water supplies is very small. Fertilisers may have to be used to maintain the desired growth rate, particularly if they were not used at planting.

The thicket stage

Once the trees have closed canopy and suppressed the competing

vegetation completely the forester leaves his plantation alone
until the time comes for the first thinning. If however
inspections suggest that the rate of height growth is less
than it should be, and this is found to be due to a deficiency
of nutrients, then aerial fertilisation will follow. As with
applications after ploughing, some fertiliser will inevitably
fall into drains, but concentrations of the materials in the
water are likely to be lower than in the first instance
because the shallow furrows would have grassed over and may
have been crossed by tree roots, the ridges will be covered
by trees or vegetation, and the catchment will be less flashy
than immediately after ploughing.

The pole or thining stage

Interventions to remove a proportion of the crop will be made
on a cycle of a number of years, which may be anything from
three in the fastest growing crops to 10 in slow growing ones.
Indeed, in exposed areas on thin soils where trees are likely
to blow over, or where remoteness and haulage costs reduce
profitability, there may be no thining at all. Up to the
early 1960s trees were planted at between 1.37 and 1.5 m
apart and these plantations are now embarrassing because of
the high cost of removing low value produce in order to give
the remaining trees room to grow. The practice of line
thinning, that is the removal of every third or fourth row
of trees, which is cheap and easy, has therefore been adopted.
This is followed at the next thinning by a herring-bone or
chevron pattern based on the missing line of trees, and only
at the third and subsequent thinnings will individual trees
be selected. Sometimes the line and chevron thinning are
combined in one operation, with selective thinning thereafter. In contrast current planting practice is to space at
about 2 m and this means that the first thinning will be
delayed, as will the hydrological impact of the growing crop,
but will be more profitable.
  Each thinning reduces water uptake, interception and
transpiration, but because in any forest some part is thinned
every year, the overall effect on the water yield is hidden.
The water yield will however be higher in a thinned forest
than in an unthinnned forest (Holstener-Jørgensen, 1967).
  Maintenance of drains has been emphasised in the past but
the lack of any suitable machine has made it difficult to
justify with present costs. It is done to ensure that root
systems developed under the influence of running drains will
not be killed by waterlogging when the drains block up. The
important point hydrologically is that keeping primary drains
open will maintain the flashy characteristic imposed when they
were first dug, while allowing them to fill up will smooth
the hydrograph. Figures 5 A and B illustrate the two extremes.

Clear felling

The universal pattern in British upland forests is to clear
fell at maturity, which will be at about 50 years in the best
Sitka spruce to 70 years in poor Scots pine. In fact, many

Figure 5. A. Drains in a Sitka spruce plantation aged 32
     years. The drains were dug at planting. Tree roots have
     been allowed to cross the drains, which have filled up
     with branches and litter. (Forestry Commission Photo,
     D4910.)
     B. Drains in a 32 year old Sitka spruce plantation. The
     drains were deepened to 60 cms ten years after planting
     and regularly cleaned thereafter, they will still be
     active in removing water from the forest. (Forestry
     Commission Photo, D4909.)

plantations do not stay the full course because they blow
over or because the danger of windblow is so great they are
felled before reaching the critical height. There will be
an immediate and large increase in the runoff following clear
felling and there is also likely to be an increase in suspended
solids in the runoff water, due mainly to the harvesting
machines. The effect on runoff has been best shown at Coweeta
in the U.S.A. where the clear cutting of a small catchment
was repeated after 23 years, with almost identical increases
in water yield (Hibbert, 1967). From the evidence cited above
one can infer that clear felling in Britain will bring about
the same changes as have been observed in catchment studies
elsewhere. The forest floor at clear felling will of course
be different from the pasture it replaced; there will be more
litter, surface layers will tend to be more porous and

infiltration will be more rapid, so the likelihood of sheet
runoff in heavy rainfall will be less.

The classical managed forest in western Europe was supposed
to have the same area of forest under each age class and the
same area felled each year, which with a regular thinning
programme gave a sustained yield of timber. The hydrograph
of such a forest would be steady from year to year. In
Britain however the afforestation of large upland areas has
usually been completed over relatively few years so that there
is no even distribution of age classes. Nevertheless, because
of different rates of growth on different soil types and
because of windblow, the distribution of age classes is likely
to be more regular in the second crop than in the first. A
great increase of water yield following the clear felling of
very large areas is therefore unlikely, although as a forest
planted over a short period enters the clear felling phase
there is likely to be some increase in yield.

Replanting usually follows clear cutting within a year,
or within two at the most. In the past the forester used to
wait several years, until the wave of beetles which bred in
the stumps and roots of the felled trees and whose adults fed
on the young planted trees had died down. Nowadays the trees
are protected with chemicals against these attacks. Replanting
follows as rapidly as possible, which helps to get the new
crop established before weeds become rampant and of course
shortens the rotation. Provided the original drainage system
was well constructed only minor repairs should be necessary.
However in a number of forests which have been clear felled
and are now being replanted, some form of cultivation appears
necessary, not only to give a new ridge raised above the
general ground level, but also so that the planter can see
where to plant as he picks his way through the branches from
the previous crop.

The large machines used in harvesting as well as in
recultivation will affect soil properties. Whether any damage
is caused to the remaining trees (in thinning operations) or
whether the survival and growth of the newly planted trees
is reduced, is still in doubt. What is not in doubt is that
the puddling and rutting of the soil surface results in an
increase in the turbidity of the drainage water which reduces
the clarity of the streams leaving the forest.

Increased biological activity follows the removal of the
trees. The addition of branches and needles, the cultivation
effect of the harvesting machines and the lack of any plants
to remove nutrients results in an increase of ammonium in the
soils. Nitrifying organisms, normally scarce, proliferate
and the nitrate formed is easily leached. There is thus an
increase in nitrate in the drainage water and this has
concerned conservationists, but results from Sweden (Tamm
et al., 1974) suggest that although the effects are real
enough, the concentrations of nitrate actually observed are
not high enough to be worrying. Only if extreme vegetation
control and nitrogen fertilising a few years before clear
felling are practised is there apparently any cause for
concern.

CONCLUSIONS FOR UPLAND FORESTRY PRACTICE

Afforestation of upland catchments in Britain will at first increase water yield and responsiveness of the streams, with increased turbidity and, if fertilisers are used, more nutrients in the drainage water. Road construction will also temporarily increase erosion and therefore suspended solids in the water. As the crops grow, the water yield will fall and the catchment will become less responsive than the original sheep pasture.

The primeval forest of mythology gave a steady yield of crystal clear water the year round, protected lower ground from floods, avalanche, and drought, provided shelter and fodder for man's animals, game for the chase, timber for his use, and a host of minor products for food and household. The modern managed forest in the uplands is rather different and is never left to find its own balance. Nevertheless, the overall yield of water should in time become steady and the floods from major rain storms reduced in intensity. Some of the undesirable effects of afforestation may however persist for a long time. For example Painter et al. (1974) show that the forested Severn is still bringing down nearly four times the amount of solids in the river water compared with the Wye, some years after drainage and road making have ceased. It is possible that harvesting operations may be partly responsible, but erosion in drains is also still going on.

The reduced water yield from the forest is a permanent feature, but are there any ways in which this and the other undesirable aspects of forest operations can be mitigated? Firstly and most obviously, the draining of large areas at any one time should be avoided and careful thought must be given to the probable effects of this and other soil disturbances during the planning phase. Leaving primary drains to fill up with silt, needles and twigs will help to reduce the flashiness induced by drainage, and because of the present day cost no unnecessary maintenance will be done. A soundly designed system should look after itself.

Fertiliser spreading and the use of other chemicals should be undertaken with consideration of the side effects. It might, for example, be necessary on occasion to restrict the proportion of any one catchment treated with phosphate in any one year, although the present practice is to try and treat as much land in an area as possible because of the economies of scale.

Clear felling in small rather than large coupes would minimise local changes in runoff pattern and is already done to some extent so as to soften the impact of operations on the landscape. Spruce is being increasingly used instead of pine, because of the higher timber yield, so if a difference in water use between the two genera is confirmed, this tendency to change species will correspondingly increase water yield.

It has been stated above that overall about 8 per cent of the land in Britain is covered with managed forests. These however are not evenly spread, as Figure 1 shows, and the proportion in some catchments could eventually rise as high as 50 per cent. Water yields from such areas would be substantially reduced in the long term but this has never yet

been taken into account in the acquisition of land for
planting, either by the Forestry Commission or private concerns.
Clarke and Newson (in press) suggest that losses are likely to
be greater at higher altitude where precipitation is heavier
and more frequent and where winds ventilating the canopy are
stronger. They go on to say '... it may well be necessary to
adopt a planned upper limit to afforestation if water supplies
are to be maintained as designed: for example, a limit could
be determined (modified, possibly, by practical constraints)
such that the expected economic return from the area is
maximised, subject to a given condition for the probability
that water yield falls below a stated level'. It is already
known that tree growth falls with altitude (Mayhead, 1973)
and forgoing the uppermost parts of a catchment would lose
the forester the least profitable part of his investment.
Similarly, leaving rocky knolls or the poorest soils within
the forest unplanted and leaving wider margins along water
courses would increase water yield for a small loss in timber,
a gain in amenity, and possibly an increase in profitability.

Many of the forester's options, such as heavy thinning
and a relatively short rotation, tend to increase the water
yield from the forest. But he has only limited room for
manoeuvre and the effect on water yield over the country as
a whole will be determined more by policy on land acquisition
and use than by the way the actual operations are done.

NOTE

The views expressed in this paper are those of the author and
not necessarily those of the Forestry Commission.

7

THE EFFECT OF LAND USE ON THE
HYDROLOGY OF SMALL UPLAND CATCHMENTS *

R.T. Clarke and J.S.G. McCulloch

Institute of Hydrology, Wallingford, Oxfordshire

ABSTRACT

The work of Law and Rutter is reviewed and the Institute of Hydrology's experimental catchments at Plynlimon and Coal Burn are described. For 1972-1976, evapotranspiration from the largely wooded Severn catchment at Plynlimon was 281 ± 20 mm greater than from the grass covered Wye catchment. Ditching in the Coal Burn catchment increased sediment concentrations by two orders of magnitude. The wooded Tanllwyth sub-catchment at Plynlimon produced four times more bedload than the pastureland of the Cyff basin.

BACKGROUND

Much recent work on the losses of water from coniferous forest has its origins in the meticulously executed study by Law (1956, 1957b) on the Hodder catchment, in the Yorkshire Pennines. Concerned about the absence of research on the water balance of woodland, and about possible lower water yields from forested catchments than from catchments planted with herbaceous vegetation, Law made a careful study of the water balance of a small (0.045 ha) natural lysimeter in a plantation of Sitka spruce *(Picea sitchesis)* set in a slightly larger block of woodland (area 0.24 ha). Based on measurements collected over the period 4 July 1955 to 8 July 1956, Law found that the precipitation above the forest canopy was 990 mm; of this, 609 mm reached the forest floor, from which 279 mm appeared as runoff. The total water loss was therefore 711 mm; over the remainder of the Hodder catchment, however, the total water loss was 421 mm. Law concluded that the loss of water from the forested plantation was the greater by 290 mm.

* This is a revised version of a paper presented at the 1975 Land Drainage Conference of the Water Space Amenity Commission.

Law's results attracted the attention of many research workers, and the scepticism of not a few; scepticism based upon the small size of the plantation used in his study, which it was argued, would have led to the introduction of edge effects in both radiative and aerodynamic aspects of vegetation. Rutter (1964), summarizing evidence from his own study of water relations of *Pinus sylvestris* under plantation conditions, from work reported by Deij (1956) on the Castricum lysimeters in the Netherlands, and from East Africa work by Pereira, Dagg and Hosegood (1962), observed that evaporation from forests had generally been found to lie between 0.8 and 1.0 times the Penman estimate of evaporation from an open water surface. Although, as Rutter states, Penman himself drew no such firm conclusion in his monograph *Vegetation and Hydrology* (1963a). Rutter's own evidence suggested that actual evaporation from the plantation exceeded open-water evaporation by 10-20%, a high figure by comparison with most other results except that of Law. The main causes suggested for the higher values of actual evaporation from forests were the darker colour of the vegetation, especially of conifers, leading to greater absorption of energy; and greater aerodynamic roughness.

Law's result, if valid, was clearly of the greatest import at a time when water supply undertakings were being encouraged to afforest their catchments; several industrial conurbations, such as those of the midlands, draw water supplies from the wetter west and north of the United Kingdom, where the climate is such that land use is largely restricted to a choice between softwood production on the one hand and upland pasture of relatively low productivity on the other. Law's results suggested that extensive afforestation would result in reduced water yield in a period during which domestic consumption may be expected to rise from its present level of 37 gallons per head daily to 60 gallons by the year 2000, and during which industrial water demand for cooling and processing is likely to increase significantly.

The Institute of Hydrology began a programme of research during the 1960s to test Law's findings. Initially two catchment studies were involved. In the first, two adjacent catchments on the slopes of Plynlimon, in central Wales, were intensively instrumented for the measurement of precipitation, river discharge, and soil moisture change. The Wye catchment is 1055 ha in area and is almost entirely upland pasture: the Severn catchment is 870 ha, of which slightly more than two thirds is coniferous forest, principally Sitka spruce and Norway spruce, but with an admixture of Japanese larch. This study seeks answers to two questions. First, is the mean annual loss (precipitation minus streamflow) greater for the forested Severn than for the hill pasture of the Wye, and if so how far is the difference explicable in terms of different land use? Second, does the rapidity and magnitude of response to unit depth of precipitation differ for the two catchments, and, if so, how far are the differences explicable in terms of different land use?

The second of the two catchment studies was set up on the 152 ha moorland catchment of Coal Burn, a tributary of the Irthing in Northumberland. This catchment was ploughed in 1972 and planted with coniferous forest according to standard forestry practice. Streamflow in the years following ploughing are being compared with those observed in the five years preceding it.

INSTRUMENTATION AND MEASUREMENTS: WATER BALANCE STUDIES

Accurate measurement of the components of the water balance, always a matter of the greatest difficulty, is even further complicated by the remoteness of the Plynlimon and Coal Burn catchments, whose difficulty of access and severe climates call for instruments of extreme robustness. The following instrumentation is in operation:-

(i) networks of raingauges, consisting of 20 ground-level monthly storage gauges in the Wye catchment, and 18 monthly storage gauges in the Severn. Of the latter, 11 are at canopy level or above, whilst the remainder are ground level gauges in the unforested part of the upper Severn, or at Moel Cynnedd, a clearing in the forest at which a meteorological station is also sited. The Coal Burn catchment has 4 weekly storage gauges. In addition to networks of storage gauges, the Wye, Severn and Coal Burn catchments contain three, three and one Dines rainfall recorders respectively, whilst a replicated network of Rimco raingauges with event recorders is being installed on each of the Plynlimon catchments.

(ii) stream gauging structures. Flow from the Wye is gauged by a modified Crump weir, that from the Severn by a trapezoidal flume, and that from Coal Burn by a Crump weir. Both Plynlimon catchments contain three sub-catchments, each of which is gauged by a specially-designed steep stream structure; the sub-catchments of the Wye are those of the Cyff, Gwy and Nant Iago tributaries, and those of the Severn are the Tanllwyth, Hafren and Hore. Stream stage is gauged by a Leupold-Stevens water level recorder, whilst each major catchment also has a Fischer-Porter punched paper tape recorder as a safeguard.

(iii) automatic weather stations (in addition to the meteorological station at Moel Cynnedd, mentioned above). Each automatic weather station records net radiation, total solar radiation, temperature, wet-bulb depression, wind run and wind direction. All variables are recorded on magnetic tape at five-minute intervals.

(iv) extensive networks of soil moisture access tubes (approximately thirty on each of the Wye and Severn catchments) at which soil moisture is recorded at 10 cm depths throughout the soil profile by means of neutron probe. Measurements are taken at roughly monthly intervals, whilst for particular periods and purposes, daily readings have been taken.

In addition intensive studies of particular hydrological processes, notably interception of precipitation by forest canopy and infiltration into the soil, have their own instrumentation at selected sites within the Wye and Severn catchments. Depth of snowfall, when it occurs, is measured by photogrammetric methods, supplemented by snow courses on which snow density is also measured. Water samples are also taken to assess natural water quality. These sources

supply data on field sheets, charts, punched paper tape or
magnetic tape, and data processing presents considerable
logistic and computational problems. The suite of computer
programs for processing is periodically updated as new
instruments come into service. For example, an autoprobe is
being developed for recording moisture throughout the soil
profile at frequent intervals using a magnetic tape data
logger.

INSTRUMENTATION AND MEASUREMENTS: SEDIMENT YIELD STUDIES

Much attention has been directed towards the collection of
reliable measurements of sediment output from the Cyff,
Tanllwyth and Coal Burn basins. Volumes of sediment moved
on the Severn catchment as a whole are by no means trivial;
observations suggest that an individual flood could carry
some 45 tonnes km^{-2} of bed material, whilst a sediment trap,
with volume 80 m^3, recently excavated above the Severn
trapezoidal flume to supplement a smaller trap already
existing there, was filled in one storm.

Table 1. Physical characteristics of the Cyff, Tanllwyth
and Coal Burn catchments (after Painter et al., 1974)

	Catchment		
	Cyff	Tanllwyth	Coal Burn
Area (ha)	314	88	152
Geology	Ordovician mudstone and shales		Boulder clay
Soils	Peaty podzol/peat/ boulder clay	Peaty podzol/ boulder clay	Peat/sand
Mean channel slope	0.034	0.031	0.026
Mean catchment slope	0.175	0.154	0.039

On the Cyff and Tanllwyth sub-catchments (Table 1),
bedload is collected in concrete traps (Painter et al., 1974)
and each movement is removed and divided into eighths by
dropping the load centrally over an octagonal splitter. One
eighth is weighed and a subsample is taken for particle
size analysis. Suspended sediment samples are taken at 8
hour intervals by a vacuum sampler; these are supplemented
by observations on samples collected during flood events by
a Tait-Binckley sampler.

To determine the relative contributions to sediment yield
from streambanks, gullies, forest roads, forest ditches and
local slips, the measurements shown in Table 2 are collected.

Table 2. Measurements in source areas of sediment
movement (after Painter et al., 1974)

Catchment	Source	Measurement Method	Frequency
Cyff	Streambank	Terrestrial photo-grammetry	1 monthly
		Manual cross-sections	2 monthly, flood event
	Gulley	Sampling unit	Flood events, $\frac{1}{2}$ hourly
	Slip	Sampling unit	Flood events, $\frac{1}{2}$ hourly
Tanllwyth	Streambank	Manual cross-sections	2 monthly, flood event
	Forest ditch	Sampling unit	Flood events, $\frac{1}{2}$ hourly
	Forest roads	Sampling unit	Flood events, $\frac{1}{2}$ hourly
Coal Burn	Forest ditch	Photographic survey	6 monthly

RECENT RESULTS: WATER BALANCE STUDY

Table 3 shows the annual totals of precipitation P, streamflow Q, and the differences P - Q for the Wye and Severn
catchments for the years 1970-75. Also shown for each year
is the value of $P_S - Q_S - P_W + Q_W$, representing the annual
difference between water losses from the Severn (forest), and
water losses from the Wye (upland pasture). After the first
two years, 1970-71, it was found that the possibility existed
that occasional sediment deposition might have affected the
flows measured in the Severn trapezoidal flume, and a sediment trap was constructed upstream to keep it clear; data
for the period 1972-74 are known to be clear of this
complication, and these were used to estimate streamflow
for the earlier period 1970-71. The annual streamflow
estimates, so obtained, are those shown in brackets; there
is no evidence of any statistically significant difference
between Severn streamflow as measured, and Severn streamflow
as estimated.
   The mean value of $P_S - Q_S - P_W + Q_W$ over the five years
is 286 ± 12 mm (or 260 ± 20 mm if estimated flows are used
for 1970-71); the mean value for the last four years, free
of the complication of the earlier period, is 281 ± 20 mm.
As stated earlier in this paper, Law's estimate of the
difference between losses from his forested lysimeter and
that from the Hodder catchment was 290 mm, for the year
between 4 July 1955 and 8 July 1956. Whichever figure is
taken for the mean difference in losses between Severn and
Wye, the agreement with Law's figure is truly remarkable.
It must be noted, however, that the precipitation on Law's

Table 3. Annual values of precipitation (P), streamflow (Q) and P - Q: Wye (W) and Severn (S) catchments, 1970-75 (units: mm)

Year	P		Q		P-Q		$P_S-Q_S-P_W+Q_W$
	W	S	W	S	W	S	
1970	2869	2690	2415	1963 (1991)	454	727 (699)	273 (245)
1971	1993	1948	1562	1196 (1328)	431	752 (620)	321 (189)
1972	2131	2221	1804	1567	327	654	327
1973	2605	2504	2164	1823	442	681	239
1974	2794	2848	2320	2074	474	774	300
1975	2099	2121	1643	1406	456	715	258
Mean	2415	2388	1985	1672 (1698)	431	717 (690)	286 (260)

lysimeter, for the year that he considered, was estimated as 984 mm compared with a mean annual precipitation on the Wye and Severn of about 2460 mm; on the hypothesis that the additional water loss from the forest is the result of evaporation of raindrops intercepted by the tree canopies, then it would be possible for such losses to be of similar magnitude for forests growing in regions of quite different annual precipitation. This is because the forest canopy may be regarded as a reservoir which, when full, passes excess precipitation downwards to the soil surface, whilst when precipitation ceases, evaporation of water in the canopy reservoir proceeds until it becomes empty or is recharged by further precipitation. The loss, over a year, of water from the forest canopy will therefore depend upon the depth (D) of water per unit area of catchment that can be stored in the canopy, and on the frequency F with which it is filled. If D and F are similar for mid-Wales and the Yorkshire Pennines, then the interception losses would be similar also, despite the differences in total annual precipitation.

The full report (Institute of Hydrology, 1976) elaborates on this result. The interpretation of the water balance in terms of energy availability is considered in detail because about 89% of the annual net radiation on the Severn is used to evaporate water, compared with 65% on the Wye. This finding is used to adjust results for that part of the Severn that is unforested and the report also discusses the part played by interception of water by the forest canopy.

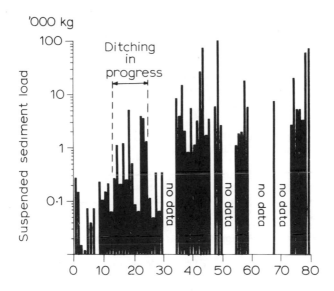

Figure 1. Coal Burn weekly suspended sediment totals
   (after Painter et al., 1974).

RECENT RESULTS: SEDIMENT YIELD STUDY

Painter et al. (1974) presented the following findings. The
most notable results to date concern the effects of ditching
on erosion and sediment transport at Coal Burn, and the
relative movements of bedload in the Cyff and Tanllwyth.
   Figure 1 gives the weekly sediment loads computed from
8-hour samples before, during and after ditching at Coal Burn.
Before ditching, sediment concentrations were typically less
than 10 ppm and even a 100-fold increase in discharge produced
a maximum concentration of only 30 ppm. Although concentrations
over 7000 ppm were recorded during ditching, low total loads
were recorded due to abnormally low rainfall which produced
only half the average runoff for the period. After ditching
was completed, successive floods have continued to transport
sediment from the basin at a rate two orders of magnitude
greater than that before ditching. It has been estimated
that, despite the rate of removal following ditching, 100 m³
of loose mineral soil was available for movement within the
drainage system at the end of the following year. Tentative
sediment rating curves have shown a large scatter even when
rising and falling stages are differentiated, which
demonstrates the limitations of lumping supply and transport
together.
   Down-cutting of ditches in the peat was limited to 10-20 mm
in the 18 months preceding March 1974. Mechanical erosion
appeared to be negligible, the main factor being flaking of
the peat as the surface dried, or during the freeze-thaw
process. Ditches in mineral material were downcut to some
50 mm in the same period, whilst severe gullying occurred on
a 30° slope, giving down-cutting to a depth of 1 m.

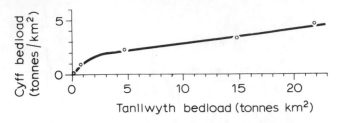

Figure 2.   Comparison of Cyff/Tanllwyth bedload transport
(after Painter et al., 1974).

    In the 9 months preceding February 1974, the basins of
the Cyff and Tanllwyth yielded 36 and 38 tonnes of bedload
respectively, in five discrete movements.   A storm with an
estimated return period of 30 years produced flood discharges
which caused both sediment traps to fill completely;   it is
estimated that a further 40-50 tonnes of bedload from each
basin was not measured.   Figure 2 gives the magnitude, per
unit area, of individual movements, showing that the forested
Tanllwyth produced nearly four times more bedload per unit
area than the upland pasture of the Cyff (42 tonnes km^{-2}
compared with 11 tonnes km^{-2}).   This was despite the fact
that both peak discharges and total volumes of flood on the
Cyff are significantly greater than for the Tanllwyth.

CONCLUSIONS

There is increasing evidence that evapotranspiration from a
forested catchment is greater than that from a catchment of
similar geology, soil and climate, used as pasture.   The
results obtained by Rutter and others, that evaporation from
forests can exceed open water evaporation, is supported by
the Plynlimon study;   the mean annual evapotranspiration
from the Severn is 21% greater than open water evaporation as
estimated by Penman's 1948 formula, or 28% greater than that
estimated by his 1956 formula.   These estimates do not take
account of any   adjustment for the unforested part of the
Severn, which might be expected to make a larger contribution
to runoff, per unit area, than the forested part.   The figures
21% and 28% quoted above may therefore be too low, and it is
of interest to record that the water loss estimated by Law,
590 mm, was 36% higher than open-water evaporation as
calculated using a (later modified) Penman formula.
    The results of the sediment yield study demonstrate also
the immediate effect of forest ditching on sedimentation.
Whilst it is known from the Tanllwyth that many ditches become
permanently inactive on maturity, a number on the steeper
slopes remain subject to considerable erosion to the bedrock;
further work is required to determine how this may be
prevented.

8

PHYSICAL EFFECTS OF RESERVOIRS ON RIVER SYSTEMS

G.E. Petts and J. Lewin

*Dorset Institute of Higher Education, Poole;
and Department of Geography, University College of Wales, Aberystwyth*

ABSTRACT

*Problems and approaches in estimating downstream effects
of reservoirs are reviewed. Alteration to the magnitude
and frequency of flows, to sediment transport characteristics and to channel forms all have practical implications
for river management in Britain.*

PROBLEMS AND APPROACHES

Since the late nineteenth century, a large number of
reservoirs have been built in the U.K. In Wales alone,
there are 14 reservoirs over 1 km^2 in area, and a further
35 of over 10 ha. Many of these were built by public or
private water authorities in upland areas, though over
the years the tendency has been to construct relatively
fewer and larger reservoirs, often now used for river
regulation rather than direct supply. In addition
reservoirs have been developed for hydroelectric power
generation. These reservoirs have a considerable physical
effect on discharge characteristics and river channels
downstream. The study of such effects is difficult
because observations may be undertaken as much as a
century after reservoir construction, and there is little
direct evidence available as to discharges, channel
geometry or sediment yields before reservoirs came into
existence. Long-term discharge data may be equally
unavailable. In these circumstances, four approaches to
estimating reservoir effects are feasible.
    First, it would be useful for future studies to be
undertaken in catchments where reservoir development is
intended, as in work by Dr. R. Hey (personal communication)
in connection with the proposed Craig Goch scheme in mid-
Wales. No published work along such lines is known, but

when case studies become available, it should be possible
to compare pre-existing physical conditions on study
reaches with ones on the same reaches after reservoir
construction, provided that representative data covering
satisfactorily comparable time-spans is obtainable for
both the calibration and the impounding phases.

Second, 'paired' or multiple catchment studies may be
undertaken, so that regulated and unregulated rivers may
be compared. A British case study of this type (Grimshaw
and Lewin, in preparation) is reviewed later. Such
studies rest on the hydrological similarity between chosen
catchments, so that observed differences in hydrology or
channel morphology can be genuinely ascribed to the
effects of upstream impoundment. In detail such identity
is virtually impossible to find but in certain cases
basins may be sufficiently similar to allow use of this
approach.

Third, it may be possible to examine conditions
downstream along a single main channel containing a
reservoir along its course. This may reveal differences
between conditions in the upstream unaffected part, and
those currently found below reservoirs. Gregory and Park
(1974), for instance, plotted and regressed channel
capacity against catchment area along the river Tone in
Somerset, which contains the Clatworthy reservoir. Channel
capacity below the dam was estimated at 54% of original
capacity, with the effects of adjustment persisting for at
least 11 km downstream.

Finally, theoretical modelling of dishcarge or sediment
yields may be attempted. For example, a number of
relationships for predicting sediment transport from flow
parameters have been derived (Graf, 1971). If discharges
and flow frequencies with and without reservoir development can be established, then differences in sediment
transport may be predicted, at least in so far as the
prediction equations provide reasonable estimates when
subjected to empirical testing.

All of these approaches have their limitations by
virtue of imperfectly understood processes and the use of
possibly unrepresentative case studies. Nevertheless, the
studies of reservoir effects in Britain so far attempted
show the major features of channel and discharge transformations to be expected. These modifications are considered under the heading of discharge, sediment yield and
channel geometry.

DISCHARGE MODIFICATIONS

The construction of a dam across a river will markedly
alter the flow regime, and observations of river flows
below reservoirs have indicated a reduction both in the
annual runoff (Gilbert and Satier, 1970), and the annual
amplitude of the discharge (Anderson, 1975). However, it
is the effect of flow regulation upon peak flows, particularly the bankfull discharge having a return period of
1.5 years for stable gravel bed river in the U.K. (Hey,
1975a), which has significance for the channel processes

downstream. The discharge at bankfull stage has been used to approximate the dominant discharge to which the mean form of rivers channels is adjusted (Ackers and Charlton, 1970).

The reduction in the magnitude and timing of flood peaks results from the absorption of flood-discharges by storage within the reservoir volume proper, and from the attenuating effect of reservoir storage above the spill-weir. The latter is related to the surface area of the reservoir, which is a surrogate for the storage available above the spillweir, the hydraulic characteristics of the spillweir, which controls the head-outflow relation, and the form of the inflow hydrograph, describing the rate of catchment response to a given storm rainfall. Downstream the lag of peak discharges routed through the reservoir may de-synchronize the mainstream and tributary peaks, further reducing flood magnitudes. The regulation of peak flows will, however, become less effective downstream from the reservoir, as the proportion of uncontrolled catchment area increases. Nevertheless, numerous accounts of the substantial reduction of peak flows may be found in the literature (De Coursey, 1975), and the reduction of the fifty-year flood by over twenty percent has been reported from both America (Huggins and Griek, 1975) and Europe (Lauterbach and Leder, 1969).

The identification of differences in the magnitude and frequency of peak flow events may be readily achieved, either by the comparison of 'inflow and outflow' or 'before and after' discharge records. However the inter-pretation of changes in the flood frequency distribution require the identification of climatic variations, primarily of storm intensity and duration, and of changes of catchment characteristics which may result in an alter-ation of the rainfall-runoff relationship. Thus the determination of the precise impact of reservoir const-ruction upon streamflow may be problematical. Nevertheless, changes in the rate of catchment response to storm rainfall may in part be related to direct precipitation on the reservoir surface, especially where the reservoir inundates more than five percent of the catchment area (I.C.E., 1975a), and to the shortening of stream lengths following the inundation of tributaries by the reservoir. More commonly an alteration of the inflow hydrograph may occur following a change in the land use of the headwaters. The affores-tation of the reservoired catchment, often undertaken to stabilize the valley side slopes and to reduce soil erosion, and hence sediment yield to the reservoir, may substantially reduce winter and spring flood magnitudes (Schneider, 1969). Moreover, a continual change in the distribution of peak flows downstream from reservoirs may be anticipated as reservoir storage capacity will contin-ually be lost through sedimentation.

The range of effects that British reservoirs may have upon the magnitude of flood discharges may be illustrated by reference to a variety of reservoired catchments, through which flood hydrographs, derived from comparable storm rainfalls and antecedent conditions have been routed (Table 1) using the N.E.R.C. (1975b) Fortran Flood Routing

Program.   The detailed changes within the flood frequency
distribution consequent upon reservoir construction may,
however, be examplified by the consideration of a single
catchment.   The River Hodder, selected for the length of
discharge record available, has been regulated since 1933
by Stocks Reservoir, controlling an area of 37.45 km^2.
The discharge record indicates that, since reservoir construction, the magnitude of the bankfull discharge has
decreased by over 25 percent and the frequency of peak flow

Table 1.   Regulation of flood discharges downstream from
selected British reservoirs.

Reservoir	Proportion of catchment inundated (%)	Spillweir length for catchment of 1km^2 (m)	Ratio of inflow-out-flow hydro-graph time to peak.	Peak flow reduction (%)
Avon Dartmoor	1.38	0.45	1.57	16
Fernworthy Dartmoor	2.80	0.46	1.86	28
Meldon Dartmoor	1.30	0.64	1.36	9
Vyrnwy mid-Wales	6.13	2.22	3.20	69
Sutton Bingham Somerset	1.90	0.43	1.41	35
Blagdon Mendips	6.84	0.49	1.69	51
Chew Magna Mendips	8.33	0.44	2.89	73
Stocks Forest of Bowland	3.70	0.91	2.09	70
Daer S. Uplands	4.33	0.67	2.21	56
Camps S. Uplands	3.13	0.61	1.82	41
Catcleugh Cheviots	2.72	0.95	2.88	71
Ladybower Peak District	1.60	1.53	2.38	42

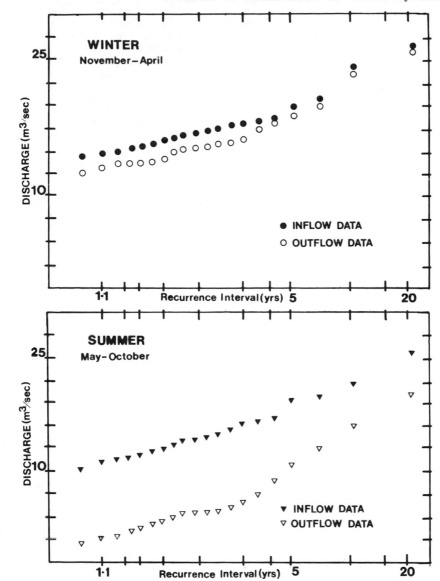

Figure 1. Seasonal variations in the alteration of the flood frequency distribution consequent upon reservoir construction on the River Hodder.

events has been markedly reduced, particularly the frequency of summer events (Figure 1). During the summer, often the period of most intense rainfalls, a storage volume may be available within the reservoir, permitting the absorption of flood discharges. During the winter, however, the reservoir may be at or near spillweir capacity so that the reduction of peak flows will be dependent upon the attenuating effect of storage above the spillweir, although the draw-down of reservoir storage may be deliberately achieved in order to increase flood control.

Examination of the annual flood frequency distribution for the River Hodder indicates that whilst the magnitude of the more frequent events will be markedly reduced, the regulation of the rarer events will be less effective. The Dartmoor Avon - a high runoff area, and the Somerset Yeo, a lowland catchment, also display a similar trend (Table 2). A given storage at a certain degree of emptiness might completely absorb a low magnitude, more frequent event, yet have an insignificant effect upon a rarer event of higher magnitude. Therefore the outflow flood frequency curve tends to be considerably below the inflow flood frequency curve for more frequent events, and to approach it more closely for the rarer events.

Table 2. The ratios of post to pre-dam discharges for flood magnitudes of selected frequency.

	Recurrence Interval (years)			
	1.5	2.3	5.0	10.0
R. Avon Avon Reservoir	0.90	0.89	0.93	1.02
R. Hodder Stocks Reservoir	0.83	0.86	0.84	0.95
R. Yeo Sutton Bingham Reservoir	0.52	0.61	0.69	0.79

Data are ratios of post to pre-dam discharges

The marked reduction of the bankfull discharge may have implications for the adjustment of the stream channel below the reservoir. Furthermore, the apparent seasonality of peak flow reduction may influence the effectiveness of channel processes (Wolman, 1959). The stability of a river system is, however, not dependent upon the magnitude and frequency distribution of discharges alone, rather it is a function of the interelationship between water discharge and sediment load.

SEDIMENTS

Reservoirs trap sediment and so provide a means of estimating catchment sediment yield (e.g. Schumm, 1954). Published work on sediment accumulation rates in British reservoirs is sparse, and has generally relied on simple ground survey techniques undertaken at low water conditions (Table 3). It is possible to make various estimates of erosion loss depending on the parts of the catchment contributing sediment, as Young (1958) observed in his pioneer study. Furthermore observed sediment accumulation depths may be difficult to interpret (Slaymaker, 1972), so that volumetric estimates may be both difficult and unreliable. Nonetheless British reservoirs are likely to be relatively efficient sediment traps and the loss of sediment load to rivers downstream has a variety of consequences.

Table 3. British reservoir sedimentation rates.

	Sedimentation rate $(m^3 \ km^{-2} \ yr^{-1})$	Time period (yr)	Source
Catcleugh, Nr. Tyne	114	55	Hall 1967
Craig Goch, mid-Wales	-	50	Slaymaker 1972
Cropston, Leicester	12	95	NERC 1976b
Deep Hayes, Stafford	19	116	NERC 1976b
Strines, Sheffield	96	87	Young 1958

Rates are given per unit area of total catchment. Sediments may in practice be derived from only part of this area, as Young (1958) emphasised.

First, a sediment load deficit may lead to channel scour (Komura and Simmons, 1967), at least where discharges past the dam spillway remain competent and where bed materials are erodible. Direct supply reservoirs in upland Britain, often sited astride bedrock channels and where high discharges are considerably reduced, may not fall into this category. A study of the Derbyshire Derwent below Ladybower reservoir showed no change until the confluence of the River Noe, 3 km downstream (Petts, 1977).

Second, the isolation of upland sediment sources may considerably reduce the sediment yield from a catchment as a whole. This may be illustrated by a comparison between suspended sediment yields of the impounded Rheidol and the Ystwyth catchments in mid-Wales (Grimshaw and Lewin, in preparation). These catchments are broadly comparable in size and general characteristics, with the Rheidol slightly the larger, rising both absolutely and on average to greater elevation, and with higher precipitation and stream frequency. A total of 54% (98 km^2) of the Rheidol catchment is affected by impounding for hydroelectric generation, and although water is not exported from the catchment, low flows are effectively augmented and high flows reduced by river regulation. Measurement of suspended sediment concentrations, and the calculation of loads using sediment rating curves and flow frequency data, revealed that in two separate years yields on the Ystwyth were 7 and 16 times those on the Rheidol (Table 4).

Table 4. Sediment loads for the River Ystwyth and Rheidol

		Suspended sediment (tonnes)	Bed Load (tonnes)
Ystwyth	1973-4	43200	4800
	1974-5	12300	1600
Rheidol	1973-4	2700	216
	1974-5	1700	

The sediment yield characteristics of the Rheidol appear to have been affected by impounding in two ways. First, discharge regulation affects the frequency with which bed sediments can be transported. Second, sources of sediment, notably suspended sediments, in the upper half of the catchment are now isolated from the lower river. Both factors, sediment supply and discharge regulation, are likely to be important below other reservoir sites, though quantitatively the results obtained for the Rheidol and Ystwyth catchments are unlikely to be repeated both because of the physical disimilarity amongst catchments and because gaugings were taken at a specific distance downstream of impounding structures which is unlikely to be duplicated. This study does seem to suggest that impounding may lead to significant alteration in sediment yields, and that the patterns of erosion and alluvial sedimentation in regulated rivers may in the long term become noticeably modified.

CHANNELS

The abstraction of sediment and the alteration of the flow regime consequent upon reservoir construction may be expected to initiate channel metamorphosis downstream (Schumm, 1969). The effect of clearwater erosion upon river channels below reservoirs has been suggested by Beckinsale (1972), but it many British upland rivers, the occurrence of coarse bed materials tends to prevent, or markedly reduce, channel degradation. A reduction in channel capacity may therefore be anticipated as a response to the reduction in the magnitude of the bankfull discharge

(Gregory and Park, 1974, 1976a).

Although Hathaway (1948) identified aggrading river channels downstream from reservoirs, relatively little attention has yet been paid to this phenomenon, possibly due to the slow rate at which changes occur. The reduction of channel capacity requires the introduction of sediment by tributaries and/or the redistribution of sediment within the river channel, and may therefore occur at a slower rate than degradation.

A change in the channel shape may also be anticipated following the alteration of the particle-size distribution of the sediment load (Schumm, 1960). River channels may adjust their width, depth, slope, roughness and planform to a change in the magnitude of the dominant or bankfull discharge. Hey (1974) has developed a hypothetical model to explain the adjustment of the hydraulic geometry of stream channels in terms of water and sediment discharge, the valley slope, and the composition of the bed and bank materials. For small channels, the stabilization of bank sediment by vegetation may also influence the direction and rate of adjustment.

Reservoir releases in the U.K. are often below the threshold of sediment transport (Hey, 1975b). So, in the absence of sediment introduction by tributaries, a simple adjustment will occur, whereby the water discharge will be accommodated within the 'pre-reservoir' channel. This results in a reduced stage for a flood of specific frequency when compared to the pre-reservoir discharge record.

The adjustment of stream channel capacity and form to an alteration of the flow regime and sediment load may be illustrated by reference to three Scottish reservoirs, regulating runoff from catchments of 3.5, 24.5 and 47.5 km^2 in area. The channels downstream from the dams actively meander within coarse sands and gravels, and enable the relatively easy identification of bankfull stage.

Utilizing the established relationship between channel form characteristics and drainage area, a common surrogate for discharge (Gregory and Park, 1974), a regional relationship was developed for the River Tweed, Daerwater, and the Water of Ae (Figure 2). Consideration of the relationship between channel cross-sectional area (CC) and drainage area (DA) reveals a marked reduction in the capacity of the channel below each of the reservoirs. However, the specific changes in the form of the river channels are not simply related to changes in the flow regime. The quantity and particle-size distribution of the sediment load, sediment introduced by tributaries, the stabilizing effect of vegetation, and the nature of the pre-existing channel are also liable to change.

The adjustments of channel form below the dams may be described by the analysis of the magnitude of change of pertinent channel form parameters (Table 5). 'Primary factors' are those channel characteristics whose dimensions are greater than two standard errors from the predicted value, based upon the regional relationship, in more than fifty-percent of the channel sites surveyed. 'Secondary factors' indicate changes within between twenty-five and

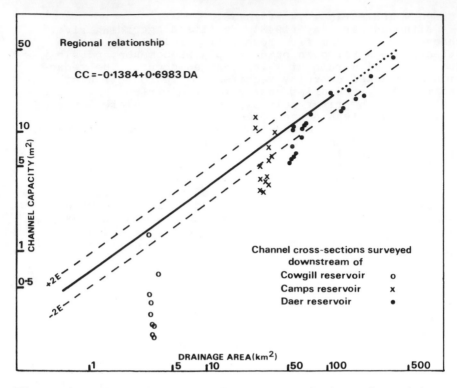

Figure 2.   Channel changes downstream of three Scottish
            reservoirs.

fifty-percent of the sites.   The 'no-change' column refers
to the observed channel parameters which are within one
standard error of the predicted value in over fifty-
percent of the cases.
      Below Cowgill reservoir the reduction of the channel
capacity is associated with the complete change in the
form of the channel.   At this small catchment area the
stabilization of sediment by vegetation is important,
resulting in the preferential reduction of channel width
to depth.   Degradation has occurred immediately below Camps
reservoir so that the channel has been enlarged by over
seventy-five percent.   However, the amount of degradation
decreases rapidly away from the dam so that within 250
metres the channel capacity has been reduced by fifty percent.   This reduction is associated primarily with a
reduction in channel width, related to the redistribution
of sediment within the meander system which is augmented
by sediment from tributaries.   The effect, however,
decreases downstream as the channel becomes increasingly
stable.
      The channel below Daer reservoir has been divided so
that the changes of channel form within reaches above and
below the confluence with Portrail Water, a major tributary
having a catchment area equal to forty-five percent of that

Table 5. Changes in channel form below Cowgill, Camps and Daer reservoirs.

	Primary factor	Secondary factor	No change
Cowgill Reservoir	Width Wetted perimeter Width-depth ratio Hydraulic radius	Mean depth	
Camps Reservoir	Width Wetted perimeter	Hydraulic Radius	Mean depth Width-depth ratio
Daer Reservoir	Mean depth		Width Wetted perimeter Width-depth ratio
Drainage area less than 100 km²		Hydraulic Radius	
Drainage area greater than 100 km²	Width	Wetted perimeter Width-depth ratio	Mean depth Hydraulic radius

of the mainstream, may be isolated. Upstream of the confluence a reduction of the channel capacity occurs immediately below the dam related to the sorting and redistribution of sediment. Although this effect decreases rapidly downstream from the dam, a reduction in channel capacity is again evident below the confluence with Portrail Water where width is the primary factor.

In response to changes in the flow regime and sediment load consequent upon reservoir construction, channels in mobile materials appear to adjust their capacities by a reduction in width. The adjustment of channel form will not, however, be uniform throughout the length of the river: rather, expansions and contractions within the pre-existing channel will permit the preferential occurrence of erosion or deposition. Furthermore, aggradation at some downstream point, say due to the deposition of sediment by a tributary into a mainstream which has lost the ability to 'flush' the material, will initiate feedback mechanisms affecting the stability of reaches upstream.

CONCLUSIONS AND MANAGEMENT IMPLICATIONS

Over the past two decades geologists (Leopold and Maddock, 1954), ecologists (Turner, 1971) and engineers (Serr, 1972) have become increasingly aware that the construction of large dams may cause major, and often adverse changes within river systems. The increased demand upon the unevenly distributed water supplies in Britain has resulted in an increase in reservoir construction so that, to-day, there are over eighty reservoirs with capacities of greater than ten million cubic metres. Furthermore, the rapid encroachment of urban development and industrialisation has necessitated the consideration of major flood control schemes for the future.

The construction of reservoirs within a river system will markedly alter the magnitude and frequency distribution of flows and significantly reduce the sediment load available to reaches below dams. The successful operation of flow regulation schemes, necessary for both flood control and water supply, requires consideration of the geomorphological and ecological consequences. The reduction in the magnitude of the bankfull discharge may result in the reduction of channel capacity which may adversely effect the efficiency of flood-water transmission downstream. Although the more frequent floods are markedly reduced, it has been demonstrated that the effect upon rarer events may be negligible.

The erosion of river channels below dams has been widely reported in the literature, but in Britain flows are often reduced below the threshold for sediment transport so that degradation is of only localised significance. However, sediment introduced by tributaries during times of flood may be deposited within the regulated main channel affecting channel slope which will initiate feedback mechanisms influencing the stability of upstream reaches. Furthermore, the lowering of tributary base level by the removal of mainstream flood peaks may result in tributary degradation, increasing the supply of sediment to the main channel.

Overall effects within catchments as a whole must depend on the proportion and nature of the catchment affected, but the study of sediment yields on the Rheidol suggested that sediment yeild effects may be extended well downstream. This is likely to alter sedimentation patterns on floodplains, notably sedimentation of fine deposits which commonly accumulate over coarser channel sediments to form higher-quality soil parent material, and in estuarine and coastal areas where patterns of sedimentation may be altered.

Both the alteration of the quantity and particle-size distribution of sediment loads, and the metamorphosis of river channels consequent upon the reduction of flood magnitudes, may affect the biota of freshwater and estuarine or marine environments. Many fish populations are dependent upon annual flooding for food, migration and spawning: the migration of salmonids, for example, may be adversely affected by regulation (Alabaster, 1970).

Engineers in Britain today are concerned with the impact of reservoir construction upon river systems, and much discussion has been concerned with the setting up of residual flows and the allowance of freshets in order to 'protect downstream interests' (I.C.E., 1972). There is enough evidence to suggest that this concern should be paralleled by one for changes in channel form and sediment characteristics, and for the physical and other effects that discharge modifications below reservoirs may have.

ACKNOWLEDGEMENTS

The authors are grateful to the Natural Environment Research Council who provided a Postgraduate Research Studentship and to the North West Water Authority, South West Water Authority, Wessex Water Authority and the Welsh National Water Development Authority for their assistance and co-operation.

9

THE HYDROLOGY OF A PEATLAND CATCHMENT IN NORTHERN IRELAND FOLLOWING CHANNEL CLEARANCE AND LAND DRAINAGE

D.Wilcock

School of Biological and Environmental Studies, New University of Ulster

ABSTRACT

The effects of an arterial channel clearance scheme on the hydrology of a small upland catchment are outlined. The catchment is covered with large areas of peat, the low permeabilities of which are thought to determine in large part the nature of the hydrological adjustments which follow the clearance scheme. Flood flows appear to be reduced in magnitude and frequency throughout the post-clearance record. Total annual water yields and flows in the intermediate and low ranges are much increased in the period immediately following clearance but these effects are not sustained. The scheme appears to be temporarily effective in withdrawing water from storage in the catchment but net annual replenishment starts within two years and current estimates suggest that restoration to initial storage conditions will take about twelve years, maybe less. It is suggested that more attention needs to be paid to improving the throughput of water in peatland drainage basins in addition to increasing capacities for output.

INTRODUCTION

Land drainage for agricultural improvement has been carried out throughout Britain for more than two hundred years and both Green (1973, 1974) and Cole (1976) have described at length the techniques of drainage developed and the changes which drainage has induced. The wide range of environmental effects produced by land drainage in different parts of the world have been admirably summarized by Hill (1976), who points out that despite the well-known general nature of these effects very little detailed evidence of specific changes has so far been presented.

In Ireland poor land drainage has always been a major obstacle to agricultural improvement. Peat in one form or

another covers 20% of the total land area, gley soils a
further 22% and peaty gleys about 5% (National Soil Survey
of Ireland, 1969). Thus nearly half of the total land
area in Ireland as a whole is covered by soils which are
very badly drained in their natural state. The history of
land drainage in Ireland has been described by Common
(1970) who presents figures to show that since 1842 more
than 4000 km^2 of agricultural land, (about 6% of the total
land area), have been improved by drainage. In Northern
Ireland, a comprehensive three-phase land drainage strategy
is currently being implemented by the Northern Ireland
Department of Agriculture. The first phase, begun in 1947,
involves widening, deepening, and straightening of the main
arterial waterways in the province. The second phase,
begun by the Department of Agriculture in 1964, involves
similar modifications to the so-called 'minor' watercourses
of the province. The third phase, responsibility for
which at the moment remains with individual farmers,
involves the installation of field drains to facilitate
removal of excess soil moisture. So far, more than 2800 km
of 'main' and 'minor' waterways have thus been improved by
clearance (Wilcock, 1977a), the total cost of all improvements completed or approved exceeding £14 million. The
effect of these schemes however has rarely been evaluated
and the present paper is a preliminary attempt to assess
the impact of a small channel clearance scheme on the
hydrology of a peatland catchment.

HYDROLOGICAL PROPERTIES AND BEHAVIOUR OF PEAT

The wide distribution of peat in Ireland necessarily means
that most catchments on which drainage schemes are undertaken frequently contain peat in one form or another.
Until recently, little was understood about the hydrology
of undrained peatlands and even less about how peatland
catchments might respond to drainage. Recent research has
highlighted several important behavioural properties of
drained peatlands and it is appropriate to review them.
    Dooge (1975), summarizing data from several sources,
shows that most hydrological properties of peat, in particular storage coefficients and hydraulic conductivity,
depend on the degree of decomposition. A highly decomposed
peat at saturation will not retain as much water as a poorly
decomposed peat. With increasing soil suction however the
position is steadily reversed and at field capacity decomposed peats may hold twice as much moisture as undecomposed
peats (Boelter, 1964, 1969). Hydraulic conductivities also
vary with decomposition. Field values as high as 3.3 x
10^3 cm/day for undecomposed peats and as low as 1 cm/day
in decomposed peats have been obtained by Boelter (1965).
In Britain field values as low as 8 x 10^{-3} cm/day have been
obtained by Ingram (1967) for a mire expanse. Other
studies, notably those of Bay (1967), Sturges (1968),
Burke (1969, 1975), and Baden and Eggelsmann (1970) confirm
these findings.
    Water-table changes following peatland drainage are also
found to depend on peat decomposition. Well-decomposed

peats on the west coast of Ireland with Van Post humification values as high as H5 at the surface (Van Post and
Granlund, 1926; Davies, 1944) and reaching H10 at a depth
of 3 meters have been studied by Burke (1975). Here, the
installation of drainage ditches one-metre deep was
effective in lowering the water-table only within a distance of 1.83 m from each ditch. Boelter (1975) confirmed
Burke's findings for well-decomposed peats but showed that
in less-decomposed peats water tables could be lowered
quite easily with a much wider spacing of open ditches than
Burke found necessary. Rayment and Cooper (1968) also
report considerable lowering of a water-table up to eleven
metres away from a drainage ditch on a sedge-bog.
     A somewhat different problem concerns the effect of
peatland drainage schemes on the hydrographs of neighbouring rivers or streams. Several workers have reported that
drainage increases the storage capacity above the water-
table and thus ensures a less rapid response to rainfall in
drained than in undrained areas. Burke (1969, 1975),
Baden and Eggelsmann (1970) and Eggelsmann (1975) all
show that this damping down of flood peaks is indeed a
major effect of peatland drainage. This view is not
supported by the work of Conway and Miller (1960) who
showed that discharge hydrographs on a burned and freely
drained blanket peat bog in the southern Pennines tended
to be more 'flashy' than those on a neighbouring but poorly
drained bog.
     The work of Conway and Miller and of Burke stand out in
the British context as examples of work done on the hydrological effects of peatland drainage. Other workers in
Britain, notably Chapman (1965), Tallis (1973), Rycroft,
Williams, and Ingram (1975) have tended to concentrate on
the hydrological characteristics of peatlands in their
natural state. All this research provides important in-
sights into the likely effects of an arterial drainage
scheme on a peatland catchment but none of it has been
specifically concerned with this particular problem.
     The rationale behind clearance as a drainage strategy
is rarely stated but can be summarized as follows.
Increased channel capacity, it is argued, removes all flows
more rapidly from the catchment than was possible
previously. A net removal of water from storage within the
catchment takes place, water-tables are lowered, drainage
improved, and groundwater storage potential in subsequent
periods of heavy or prolonged rainfall is increased. In
post-clearance conditions all but the very largest floods
are accommodated in the new channel and overbank flooding
is therefore less frequent. Groundwater flows, it is often
claimed, are augmented.
     The effectiveness of an arterial drainage scheme in
augmenting very low flows was confirmed by Corish (1971)
although he found from his work in County Tipperary that
intermediate flows tended to be diminished. Other
questions however remain. How quickly does a peatland
catchment respond to a channel clearance scheme? How long
do the benefits last? By how much are water-tables in
peat lowered following clearance? The present paper is an
attempt to provide some preliminary answers to these questions.

STUDY CATCHMENT

The Glenullin Basin (Figure 1) lies at the western edge of
the basalt plateau in north-eastern Ireland and is 15 km^2
in area. Basalt in fact underlies most of the catchment
but is covered by a thick mantle of boulder clay. The
National Soil Survey of Ireland (1969) classifies climatic
peat and gleys as the two major soils of the basin.

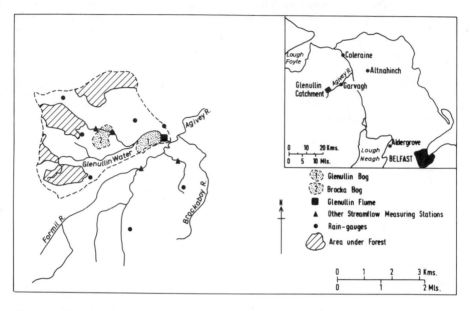

Figure 1. Location of the Glenullin basin and the distribution of rainfall and streamflow monitoring stations.

There are also some acidic brown earths. The low permeabilities of the soils, drift, and solid geology of the area
suggest that little seepage out of the catchment takes
place. Much of the blanket bog covering the uplands has
been afforested, though none of this took place during or
immediately before the study period. Forest now covers
26% of the basin. The remainder of the catchment is
permanent pasture or rough pasture. Sheep rearing and
dairying are the major occupations.
    Brocka bog, one of two raised bogs in the low-lying
central parts of the catchment is very degraded and has
been extensively cut during the past century. Glenullin
bog is a classic and essentially untouched raised bog and
has a typical raised bog vegetation with *Calluna vulgaris,
Sphagnum* spp, *Eriophorum angustifolia,* and frequent *Erica
tetralix*. In the most degraded parts of Brocka bog *Juncus
effusus* and *Eriophorum angustifolia* are the most abundant
species. Borings in the centre of Glenullin bog reached
4.75 metres before encountering grey clay. In Brocka bog

borings reached a maximum depth of only 3 metres.

Decomposition levels have been examined in Glenullin bog and do not seem to be as high as those reported by Burke (1975). At the surface Van Post humification values were as low as H2-H3 and at depths of 4 metres were only in the H9-H10 range. Wet sieving of peat samples taken from a four-metre core showed that 67% of organic matter in the section as a whole has a fibre length exceeding 0.125 mm. This would probably correspond with Boelter's hemic category, signifying intermediate levels of decomposition. (Boelter, 1969, 1972). More work on decomposition levels is currently being undertaken on all the peats in the catchment.

In the spring of 1971 a water-level recorder and rectangular pre-fabricated steel flume were installed at the outlet of the catchment. The steel flume, recessed into the bed and banks, is 2.44 metres wide, 1.83 metres long and 1.52 metres deep and was designed to provide a stable current-metering section. The rating-curve of stage against discharge derived from current meter measurements made at the flume is the basis of all streamflow calculations and has remained stable throughout the five-year period studied. The post-clearance rating curve is however very different from the pre-clearance rating curve (Wilcock, 1977b), the gauging section at all stages now carrying larger volumes of water than before clearance.

Rainfall is measured by 5 standard Meteorological Office raingauges and by a Normalair-Garrett autographic rainfall recorder. The Thiessen method of estimating areal rainfall has been adopted throughout the work.

Evapotranspiration figures used throughout the study refer to potential evapotranspiration (PE) and not to actual evapotranspiration. Two sets of PE figures are used. The first is that officially derived by the Northern Ireland Meteorological Office for a station at Altnahinch in County Antrim, using the formula described by Grindley (1970). Although well to the east of the study area this station is thought to closely approximate conditions in Glenullin. The Meteorological Office estimates of PE for Altnahinch have been adjusted, in the light of recent lysimeter experiments, to provide a second set of PE estimates which are also used in the present study. Penman estimates and modified Penman estimates for Altnahinch are presented later.

Since the autumn of 1976, 14 plastic tubes up to 1.5 metres in length and 7.5 cm in diameter have been installed in the two raised bogs to observe water-table levels. Data from these wells, though not relating directly to the five hydrological years examined in the present study, can nevertheless be used to check the feasibility of some of the changes in water storage calculated for the five year period as a whole.

In the autumn of 1971 a channel clearance scheme was completed on a one-mile stretch of the Glenullin extending upstream from its junction with the Agivey. Bushes were removed and sediment scoured from the bed and banks. The stretch of flowing water directly affected by clearance skirts Glenullin bog along its northern and eastern edges.

Man's impact on the hydrological cycle

In the five years that have elapsed since the scheme the
only other major modification to the catchment has taken
place on Brocka bog where in 1972 approximately 1800 m of
open ditches and 1750 m of field drains were installed.
The analyses which follow describe the changing hydrological behaviour of the catchment in the five years following
these schemes.

ANNUAL AND SEASONAL WATER YIELDS

Changes in water yield throughout the five year period are
shown in Table 1 where monthly and annual streamflow data
are presented as percentages of precipitation. It is

Table 1. Monthly and annual streamflow as a percentage
of precipitation.

Year	O	N	D	J	F	M	A	M	J	J	A	S	Total
1971-72	58	94	71	123	101	86	85	80	44	35	58	19	79
1972-73	28	107	104	93	99	43	62	41	8	44	27	5	65
1973-74	37	87	62	76	67	59	65	39	13	45	19	76	58
1974-75	64	76	71	80	53	52	55	42	16	22	20	25	54
1975-76	83	56	61	74	71	26	43	45	47	36	47	44	54

apparent from the annual figures that runoff rates in
1971-72 were abnormally high and have subsequently tended to
decline steadily toward an apparent condition of stability
in which annual streamflow now accounts for approximately
54% of precipitation. These figures strongly suggest that
the high flows of 1971-72 and to a lesser extent those of
1972-73 were artificially induced by the channel clearance
and peatland drainage schemes and involved withdrawals of

Table 2. Frequencies of high daily flows in winter.

Water Year	Rainfall Oct-Mar (mm)	No. of days (Oct-Mar) with daily flows greater than		
		4.0 m^3 sec^{-1}	3.0 m^3 sec^{-1}	2.0 m^3 sec^{-1}
1971-72	654	0	5	13
1972-73	687	3	6	13
1973-74	814	1	2	11
1974-75	665	0	0	7
1975-76	619	0	1	3

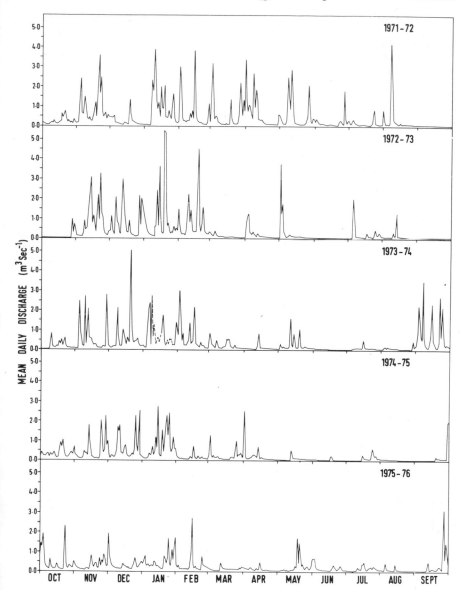

Figure 2. Mean daily flow hydrographs at the Glenullin flume.

water from groundwater and/or soil moisture storage. The
quite rapid stabilizing of annual streamflow at a relatively
fixed percentage of annual precipitation is particularly
apparent in water years 1973-74 and 1974-75, one of which
was 25% wetter than the other.

The annual stability over the last three years of record
is not however reflected by any stability in the monthly
figures. Over this period streamflow as a percentage of
precipitation has been steadily increasing in October, May,

June, and August and steadily decreasing in November, March
and April. The winter months of December, January and
February have been most stable, which is what one would
expect with generally high water tables and low evaporation
rates, whilst July and September have been most variable.

EXTREME FLOWS

The daily distribution of flows throughout the five-year
period following clearance is shown in Figure 2 and it is
apparent that the incidence of large flows has steadily
declined since 1971. The effective decrease in winter
flooding is most apparent from the data relating to water
year 1973-74 (Table 2). In the six months from October
1973 to March 1974 rainfall totalled 814 mm, almost 20%
greater than in any similar period of the five years
examined. Winter floods however were less frequent in this
period than in either of the two preceeding winters, and in
the final two years of record the decline appears to have
continued.

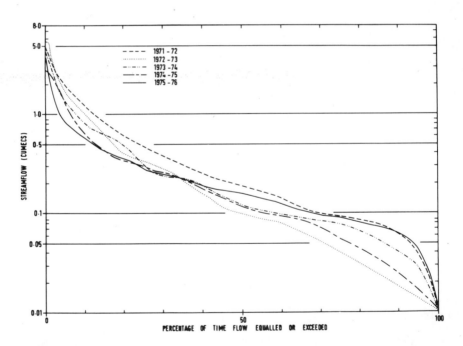

Figure 3. Flow duration curves for the years of study.

The decline in winter flooding frequency has not however
been accompanied by any apparently lasting augmentation of
low and intermediate flows in the years following clearance.
In the summer of 1972 low flows were higher than in any
subsequent year (Figure 2). But by 1972-73 the flow

equalled or exceeded 50% of the time had been reduced to half its 1971-72 value (Figure 3).even though annual rainfall had fallen by only 3% in comparison with the preceding year, and summer rainfall (April-September) by only 12%. Some recovery took place subsequently and by 1975-1976 the 50% duration flow was 70% of its 1971-72 value. It is nonetheless apparent that over the five year period there is no evidence to suggest that any sustained increase in the amount of perennial storage available to increase low and intermediate summer flows.

THE WATER BALANCE

Water balance data for the five year period as a whole and for the individual years following clearance are shown in Table 3. If residual values in Table 3 are regarded as primarily reflecting changes in storage, the periods of withdrawal and subsequent replenishment following clearance are evident. Clearly the negative residual in 1971-72 is primarily the result of increased streamflow in that year

Table 3. Water balance data (mm) for the period October 1971 - September 1976.

Period	Precipitation (P)	Streamflow (Q)	Potential Evapo-transpiration (PE)	Residual (R) = (P-(Q + PE))	$R/_P$ x 100
1971-76	5919	3684	1982	+253	4.3
1971-72	1201	945	390	-134	11.2
1972-73	1169	762	388	+ 19	1.6
1973-74	1376	799	390	+187	13.6
1974-75	1078	583	404	+ 91	8.4
1975-76	1095	595	410	+ 90	8.2

and is not due to any obvious peculiarities of the annual rainfall or evapotranspiration. Water balance data for individual months are presented in Table 4, and accumulated monthly residuals showing changes in the water balance over the five year period are shown graphically in Figure 4. The most obvious feature of this graph is the disruption to the seasonal pattern of hydrological events in the period immediately following clearance. Winter months are normally periods of storage replensihment but withdrawals from storage in 1972 appear to have started as early as January and to have continued more or less uninterrupted for a further eighteen months. Equally apparent from Figure 4 is the short-lived nature of this disruption. Although there does not appear to have been any obvious period of recharge in 1972-73, the familiar sequence of additions to storage

Table 4.   Water balance data (mm) for the period
October 1971 - September 1976.

Water Year	Month	P	Q	PE	R	PE*	R*
1971-72	October	76	44	16	+16	24	+ 8
	November	168	158	2	+ 8	2	+ 8
	December	65	46	9	+10	19	0
	January	128	158	5	-35	25	-55
	February	102	103	10	-11	24	-25
	March	115	99	21	- 5	38	-22
	April	122	104	43	-25	48	-30
	May	128	102	65	-39	68	-42
	June	79	35	71	-27	66	-22
	July	86	30	69	-13	72	-16
	August	105	61	50	- 6	57	-13
	September	27	5	29	- 7	36	-14
1972-73	October	60	17	17	+26	25	+18
	November	134	143	3	-12	3	-12
	December	131	136	2	- 7	12	-17
	January	184	171	3	+10	23	-10
	February	131	130	6	- 5	20	-19
	March	47	20	27	0	44	-17
	April	61	38	41	-18	46	-23
	May	80	33	61	-14	64	-17
	June	67	5	75	-13	70	- 8
	July	94	41	66	-13	69	-16
	August	86	23	54	+ 9	61	+ 2
	September	94	5	33	+56	40	+49
1973-74	October	83	31	12	+40	20	+32
	November	120	104	4	+12	4	+12
	December	175	108	2	+65	12	+55
	January	233	178	9	+46	29	+26
	February	138	93	8	+37	22	+23
	March	65	37	25	+ 3	42	-14
	April	26	17	50	-41	55	-56
	May	110	43	64	+ 3	67	0
	June	54	7	74	-27	69	-22
	July	95	21	58	+16	61	+13
	August	88	17	55	+16	62	+ 9
	September	189	143	29	+17	36	+10
1974-75	October	102	65	11	+26	19	+18
	November	113	86	2	+25	2	+25
	December	151	107	5	+39	15	+29
	January	167	134	5	+28	25	+ 8
	February	47	25	5	+17	19	+ 3
	March	85	44	20	+21	37	+ 4
	April	75	41	34	0	39	- 5
	May	36	15	77	-56	80	-59
	June	50	8	81	-39	76	-34
	July	91	20	72	- 1	75	- 4
	August	40	8	60	-28	67	-35
	September	121	30	32	+59	39	+52

Table 4 continued.

Water Year	Month	P	Q	PE	R	PE*	R*
1975-76	October	99	82	17	0	25	- 8
	November	97	54	3	+40	3	+40
	December	100	61	0	+39	10	+29
	January	151	112	2	+37	22	+17
	February	76	54	9	+13	23	- 1
	March	96	25	20	+51	37	+34
	April	47	20	48	-21	53	-26
	May	122	55	63	+ 4	66	+ 1
	June	70	33	69	-32	64	-27
	July	76	27	70	-21	73	-24
	August	34	16	77	-59	84	-66
	September	127	56	32	+39	39	+32

PE*   Meteorological Office estimates adjusted to take account of
      lysimeter measurements made at Coleraine.

R*   Residuals based on adjusted Meteorological Office estimates
      of PE.

in winter and withdrawals from storage in summer appears
to have been re-established by 1973-74 and to have been
maintained since.   Storage levels existing within the basin
before clearance seem to have been re-established by
December 1973 since when the basin as a whole appears to
have been getting progressively wetter.

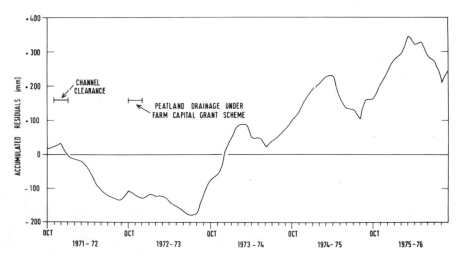

Figure 4.   Cumulative residual values plotted against time
    for the five-year study period.

This interpretation of the storage changes, using only
the first four years' data following clearance, has been
presented elsewhere (Wilcock, 1977b). When the 1975-76
data are added to those for earlier years the progressive
additions to storage on the scale shown begin to appear
somewhat unrealistic and the possibility exists that the
data relating to one of the variables used in the water
balance calculations may be inaccurate. Because it is not
measured locally, PE at once appears likely to be a source
of some error. Another reason for suspecting PE is that
the applicability of the Meteorological Office PE formula
to Irish conditions has not been precisely determined. It
has been reported for example (Morgan, 1962) that mean
annual PE as measured by lysimeter at Valencia in south-
western Ireland is 8.2% lower than the PE value predicted
by the Penman formula (1950). Until recently no similar
experiment had been undertaken in Northern Ireland, but in
1975 two grass lysimeter experiments of the type described
by Green (1959), Ward (1963), and Morgan (1962) were
started at the New University of Ulster campus in
Coleraine and at Aldergrove, County Antrim. The latter
experiment is being conducted by the Northern Ireland
Meteorological Office. Initial results from both these
experiments suggest that PE estimates currently being made
by the Northern Ireland Meteorological Office and based
essentially on Penman's formula (1963a) underestimate
annual evapotranspiration. The Coleraine results point to
a serious underestimate in winter. The Aldergrove results
suggest large underestimates in summer. The monthly
difference between Meteorological Office and lysimeter PE
estimates at Coleraine have been added to the Penman PE
estimates made for Altnahinch to see what picture emerges
of water balance events following clearance. The lysimeter-
amended PE estimates for Altnahinch are presented in
Table 4.

The sensitivity of water balance studies to different
methods of measuring precipitation and evapotranspiration

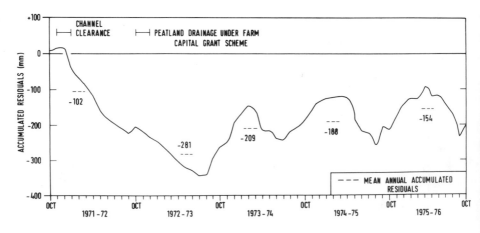

Figure 5. Cumulative residual values, derived using
Coleraine lysimeter-corrected estimates of PE.

is already well-known.(Edwards and Rodda, 1972). In the
present study, the graph of water balance storage changes
derived using lysimeter-amended PE values (Figure 5) produce a much less dramatic sequence of events than is shown
in Figure 4. The period of withdrawal from storage in
1971-72 and in 1972-73 is still apparent and indeed is more
severe owing to the higher winter PE estimated derived from
the Coleraine lysimeter. Recovery commences in the autumn
of 1973 but subsequent summer withdrawals in the following
three years almost compensate each period of water recharge
and an approximate seasonal balance appears to be re-
established as early as 1973-74. Mean monthly accumulated
residuals for each year are shown in Figure 5 and successive annual additions to storage do appear to be taking
place even though the original volume of water withdrawn
from storage does not yet appear to have been replenished.
If future annual additions to storage continue to be of the
same magnitude as those suggested on Figure 5, the process
of recharge to the zero datum line will be completed in
another 6 years, some 10-12 years after initial clearance.

If the Aldergrove lysimeter data are used to amend the
Altnahinch data, the accumulated monthly water balance
residuals follow a path intermediate to those already
described. The seasonal sequence of recharge and withdrawal is all but destroyed in the first two years. The
lowest accumulated residual (-258 mm) is recorded in
August 1973. Recovery however is rapid and zero datum is
crossed as early as February 1973. Mean annual accumulated
residuals are positive in the last two years of record
(Table 5).

Table 5. Mean monthly accumulated residuals (mm)
derived from different estimates of PE at Altnahinch.

	1971-72	1972-73	1973-74	1974-75	1975-76
Met. Office PE	-47	-130	+41	+172	+277
Met. Office PE modified by Coleraine lysimeter	-102	-281	-209	-188	-154
Met. Office PE modified by Aldergrove lysimeter	-46	-183	-57	+32	+95

WATER-TABLE MEASUREMENTS

Water-table levels measured in Glenullin and Brocka bogs in
the autumn of 1976 suggest that the sequence of water balance events described using modifications of PE based on
the Coleraine lysimeter data may probably be the most
realistic. Burke (1975) points out that water-table levels

within the well-decomposed blanket bog peats of the west coast of Ireland are typically about 0.10 m deep in winter and 0.25 m deep in summer. In the less decomposed peats of the Glenullin basin, it can be assumed that the normal summer water-table depth is somewhat lower, at about 0.3 m. If therefore zero datum in Figures 4 and 5 is taken to represent normal September storage conditions in the years preceding clearance and if a storage coefficient of about 0.5 is ascribed to the Glenullin peats, then the accumulated negative residual of 0.2 m in September 1976 (Figure 5) represents a water-table depth of 0.7 m, about 0.4 m below normal. Actual water-tables in Glenullin bog at the beginning of October 1976 ranged from 0.125 m to 1.12 m and had an average depth in four wells of 0.56 m. The wells had been installed one week at the time these measurements were recorded. In Brocka bog midway through October water-table levels in four wells were 0.1 m, 1.04 m, 0.63 m, and 0.61 m., the average depth here being 0.57 m. The mean water-table depth in both bogs was thus about 0.13 m higher than the 0.7 m depth crudely estimated from the earlier assumptions, a difference which represents a greater level of agreement between observed and calculated water-table levels than can be achieved using estimates based on the two alternative water balances. The accumulated residual in September 1976 derived using unmodified PE data from Altnahinch is +263 mm. This would require a water-table in 1976 some 0.53 m higher than in September 1971, a difference which could be satisfied only if the water-table in September 1976 was at the peat surface or if in September 1971 it had been at a depth of 1.10 m. The first alternative was manifestly not the case and the second seems improbable. Similar reasoning applied to the water balance as derived using Aldergrove modifications of PE would require water-table depths in September 1976 of only 0.11 m or, alternatively, depths in September 1971 of 0.77 m. Again the measurements made in September 1976 do not come close to the first of these alternatives and the second still represents a very low value for normal conditions in undrained peat.

CONCLUSIONS FOR MANAGEMENT

The Glenullin catchment is not very representative of the drainage basins on which arterial improvement schemes have so far been carried out in Northern Ireland. It is in an upland area whereas most schemes are carried out in lowland catchments, and the clearance operations were relatively small-scale. The absence of pre-clearance records further complicates the problems of accurately assessing the total impact of the scheme. For these reasons it would be unwise to derive a set of firm management proposals from any of the work undertaken so far. Nevertheless some preliminary conclusions do emerge and can be summarized as follows.
    1. Channel clearance, unaccompanied by any other form of drainage improvement, appears to be quite effective in removing substantial amounts of water from storage and in

reducing the magnitude and frequency of higher flows, particularly in winter.

2. Low flows in summer months are dramatically increased in the short-term. There is however no evidence to suggest any lasting augmentation of summer low flows as a direct consequence of channel clearance.

3. With respect to annual yields the impact of clearance on the hydrological system seems to last only a short time and annual streamflow is stabilized as a relatively fixed percentage of annual precipitation within two years of clearance.

4. Storage replenishment appears to re-commence about two years after clearance, but there is doubt about the rate at which it occurs. The most likely estimate suggests that the benefits of increased storage within the basin last between ten and twelve years. The shortest estimate suggests an effective storage increase lasting only two years. The former interpretation appears more consistent with recently-measured water-table levels and also with the observations that the decrease in flood magnitude and frequency is being sustained but not accompanied by any lasting augmentation of low flows. Evidently the storage created following clearance is still partly available to damp down floods but is not being maintained by ground-water outflow.

These conclusions require more elaborate testing. Work on water-table changes throughout the catchment is now proceeding. In addition work has been started measuring evapotranspiration directly from the peat. A small bottom-less lysimeter of the type described by Bay (1966) has recently been installed in Glenullin bog and it is hoped that data from this instrument will answer some of the very important questions relating to actual evapotranspiration from peat, and make possible further refinements of the water-balances described in the present paper.

Clearly the general conclusion which emerges is that the permeabilities and hydraulic conductivities of the organic and gley soils of the type found in the Glenullin basin are so low that attention must be given to improving the through-put of water in the basin as well as the output of water from it. Improvements designed to increase the rate of output alone are not comprehensive solutions to land drainage problems and, if the initial benefits of clearance schemes are to be sustained, under-drainage schemes should be implemented as soon as possible following channel clearance.

ACKNOWLEDGEMENTS

I would like to thank Dr. R. Ward (University of Hull) and Professor P.J. Newbould (New University of Ulster) for their helpful comments on the work described in the paper. I should also like to thank engineers from the Northern Ireland Department of Finance for advice and help with the streamflow measurements, and the technical staff of the New University of Ulster for help with the design and installation of field equipment and the collection of field data.

10

FLOODS ON MODIFIED FLOODPLAINS

J. Lewin, R.L. Collin and D.A. Hughes

Department of Geography, University College of Wales, Aberystwyth

ABSTRACT

*Floodplain inundation processes are reviewed and illus-
trated with respect to the River Teifi. Such processes
have been much modified by human activities. Flood
observations of the Teifi show the role of ditching in
flood modification, whilst studies of the Sence, undertaken primarily to develop photogrammetric methods for
flood mapping, illustrate the effects of communication
lines and flood storage modification. The major effects
of modification to floodplains are qualitatively
summarized.*

INTRODUCTION

It is generally well known that British floodplains have
been progressively developed for urban, industrial and
other purposes and that this has increased the potential
for flood damage (Nixon, 1963; Addyman in Newson, 1975;
Penning-Rowsell and Parker, 1974). What may be less
generally appreciated is that such developments may themselves have produced feedback effects on flood discharge
characteristics. Whilst extensive flood protection and
embankment schemes may solve flood problems locally they
may also, if only in a limited way, create other problems
downstream as a result of decreased flood storage. The
great majority of British floodplains have been modified
by human activity in some form or other, and it seems
useful in an exploratory way to consider what effect this
might have.

This paper has developed following two pieces of work.
The first was an attempt to monitor what happens to
flooding waters and patterns of inundation during a flood,
and the second, an effort at flood-extent mapping
following research being undertaken in collaboration with

the Severn-Trent Water Authority. Thus observations have been carried out in the Teifi in mid-Wales and the Sence, a right-bank tributary of the Soar just to the south of Leicester. Observations in both areas suggest that progressive human modification of floodplain environments is likely to have altered flood charac- teristics in a number of significant respects.

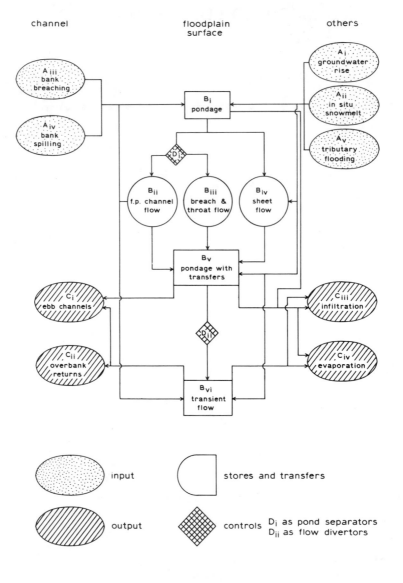

Figure 1. An inundation model (after Lewin and Hughes, in preparation).

INUNDATION PROCESSES

Processes involved in the movement of water between river channels and floodplains are summarized in Figure 1 (Lewin and Hughes, in preparation). Inundating waters may come not only from the stream channel but also from snow melt (Popov and Gavin, 1970) and groundwater rise. Water from the channel may come from discrete bank breaches in levees (Richter, 1965) or where abandoned cut-offs or dead sloughs are still open at one end. It should be appreciated that levee crevasses and cut-offs are normal floodplain features and that deposits associated with them are common in alluvial sediments (Allen, 1970). Flow through such breaches is essentially a process which can occur before bankfull stage is reached, and the overflow of waters along great lengths of the bank (bank-spilling) may only occur after a significant proportion of the floodplain contains water. Bank-spilling is also a process affected critically by the geometry of the floodplain, with waters flowing downslope away from channels or backing-up against a rising surface gradient (Schmudde, 1963).

Flows on floodplains may be via ephemerally-occupied channels, as shallow sheet flows, or via throats or breaches in natural or artificial elevations between relatively still-water pondage areas. Such waters may then be returned to channels downstream via ebb channels or as overbank flows. Alternatively water may be lost through infiltration or evaporation over a rather longer timescale.

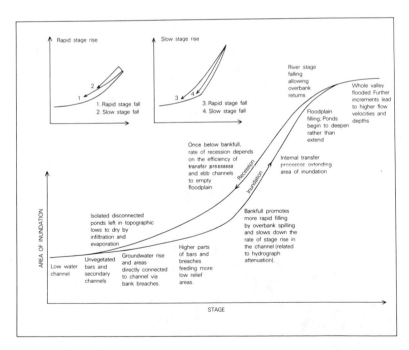

Figure 2.   Relative flood stage versus area of inundation (after Lewin and Hughes, in preparation)

Figure 3.
A floodplain
breach showing
flow at three
stages of
inundation.

These processes may operate in a particular sequence of inundation and recession (Figure 2) (Lewin and Hughes, in preparation). Thus flooding may start with groundwater ponding which in Britain may be extensively present under winter conditions without overbank floods. Breach and overbank flow may then take place to be followed by the development of pondage areas which eventually link up to form a transient current across the whole floodplain. As flood-stage is lowered, recession may be marked by the increasing dominance of ebb-flow channels, the shrinking of storage areas, and the eventual drying-out of the floodplain.

These processes may be illustrated by observations from the River Teifi. Figure 3 shows a floodplain depression at three flood stages : an initial thread of water develops into a high-velocity breach flow, and eventually to transient flow across much of the floodplain. Figure 4 shows an initial bank breach in the Teifi followed by ponding and the amalgamation as flows across the floodplain between depressions are established. With recession, flows in the foreground bank breach are reversed and it operates as an ebb channel.

The importance of processes of this kind in achieving a rate and volume of water transfer or storage will vary from site to site, and with individual flood conditions. Similarly, there is strictly no unique relationship between flood stage and inundation for a given site, but rather a hysteretic effect in which floodplains are likely to contain less water during stage rises than during flood recession (Figure 2). This effect depends not only on floodplain form but also on individual flood hydrograph characteristics.

FLOODPLAIN MODIFICATION

All the flood and floodplain characteristics we have so far discussed may be modified, deliberately or accidentally, by human activities on floodplains. Channel engineering may lead to both a decrease in the frequency of flood inundation and to modification in the routes whereby such inundation takes place. Structures on floodplains may lead to compartmentalization and the removal of certain areas from inundation. For example, 33% of the Rheidol and 43% of the Ystwyth floodplains in mid-Wales have been isolated to a degree from their respective rivers by the building of a Turnpike road (1812), and by railway embankments constructed in 1864, 1867 and 1904 (Lewin and Brindle, 1977). Deliberate alteration of surfaces or floodplains had been accomplished in many areas, by the infilling and cultivation of floodplain depressions, by ditching and land drainage, and by waste dumping on the margins of urban areas. This may permit subsequent land development for recreational and other purposes with a reduced risk of flooding.

Figure 4.   The Teifi at Trecefel showing three flood stages.

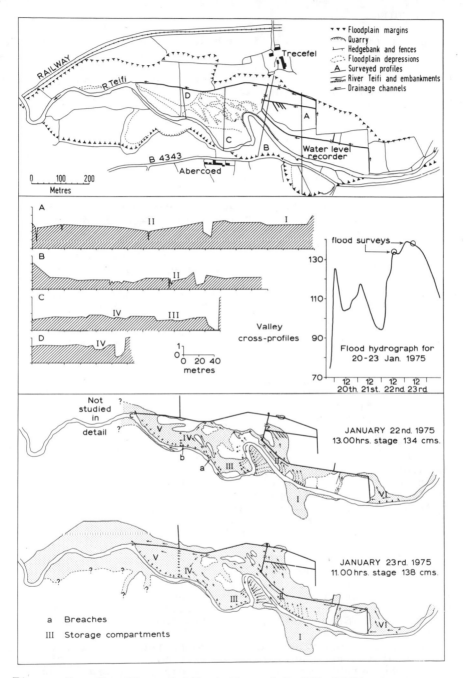

Figure 5.   The River Teifi at Treceful (SN 6758) : top, map
     showing general floodplain features; centre, floodplain
     cross-profile and flood hydrograph for 20–23 January
     1975; bottom, inundation features on January 22 and
     January 23 1975. (after Lewin and Hughes, in preparation)

TWO STUDIES

The hydrological effects of such activities may be illustrated rather more specifically with respect to reaches of
the Teifi and the Sence. The Teifi, below the raised bog
and town of Tregaron, flows between closely adjacent
valley sides. These widen to a maximum of some 350 m at
Trecefel (SN 6758) for 1.5 km before flood waters are
again confined by a railway embankment. Surveys here
indicate that the right bank floodplain slopes away from
the channel and profiles show 70% of the floodplain below
bank top height, a feature noted commonly on local rivers
(Lewin and Manton, 1975).

Two observed floods are illustrated in Figure 5.
Arrows show flow directions and shaded areas the extent of
inundation. To an extent flood processes operate quasi-
naturally as previously described, involving three flood
breaches (a-c) and a series of pondage 'compartments'
(I-VI). Groundwater pondage is not important here. The
floodplain is under permanent grassland, though natural
forest vegetation would of course affect overbank flows
considerably.

A significant feature from the present point of view
is the ditch system on the right bank of the river. The
bed level of this ditch is often 0.6m lower than the
adjacent channel bed, and in effect operates as an
additional channel, decreasing the extent of inundation
beyond it and remote from the river, decreasing pondage,
and conducting water rapidly downvalley to join the main
channel again. Inundation is also restricted by channel
and floodplain embankments, so that even in this relatively
unmodified floodplain case, flood properties are considerably modified with an increase in overbank velocities, a
decrease in pondage, and little hysteretic storage effect;
all because of the ditch system.

The River Sence provides a contrasted example. Here
the floodplain has been traversed by several railway
embankments, is paralleled by the Grand Union Canal to the
north, and urban development is encroaching onto the
'natural' floodplain area. The objective of the Sence
study is the mapping of flood extent. This would seem to
be an exercise not yet systematically undertaken by Water
Authorities, yet one of considerable importance, in practical as well as theoretical terms. It was felt that
photogrammetric methods might be the key to providing the
necessary terrain data for a predictive computer model.
Accordingly, observed flood levels and photogrammetric
ground height and plan data were obtained for a 10 km
length of the Sence Valley above its confluence with the
Soar.

An initial exercise has been completed involving the
capture of the necessary terrain information from the
aerial photographs in the form of a digital terrain model,
the calculation of flood levels for the 25 year return
period, and the prediction of a 'flood plane' between
successive observed flood positions. The intersection
between the terrain model and the flood plane may be considered to provide a close approximation to the extent of

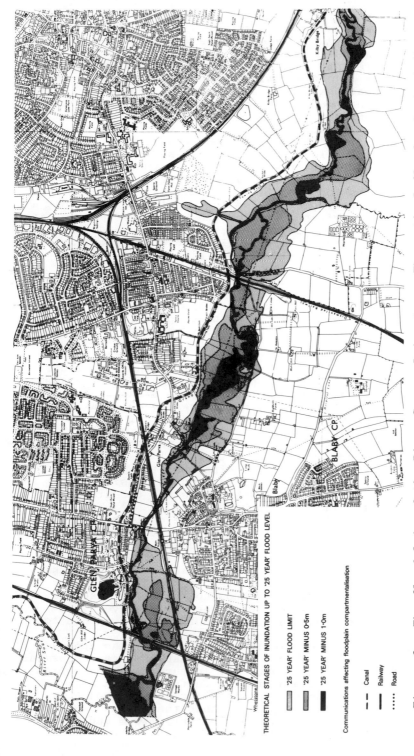

THEORETICAL STAGES OF INUNDATION UP TO '25 YEAR' FLOOD LEVEL.

'25 YEAR' FLOOD LIMIT

'25 YEAR' MINUS 0·5m

'25 YEAR' MINUS 1·0m

Communications affecting floodplain compartmentalisation

— — — Canal

——— Railway

· · · · · Road

Figure 6. The floodplain of the River Sence, Leicester. Theoretical flood extent at the '25-year' level, and water levels at 0.5 m and 1.0 m below this, are illustrated. Based upon the Ordnance Survey Map with the sanction of the Controller of Her Majesty's Stationery Office, Crown copyright reserved.

117

flood inundation for the return period in question
(Figure 6). Indeed, comparison with a field flood map
based on the same data, made by the Water Authority, shows
a very similar pattern.

At the present time, the study is still in progress,
with a number of options within the photogrammetric and
computer operations under consideration. However, there
are several more general points worthy of note. The method
assumes instantaneous level adjustment rather than a
sloping water surface away from the river. However, since
the original flood levels were measured at a variety of
positions, often along the valley sides, this is not really
an objection particularly since only maximum flood extent
is predicted. The choice of a sloping flood plane is an
over-generalisation, but is empirically justified between
successive points. It is also interesting to note that a
plane sloping surface does give the best least square fit
to the whole set of points for the valley in question.
Adjustment of areas theoretically liable to flooding, but
'sealed off' by artificial floodplain development, is
possible and desirable, though this can sometimes be
difficult from aerial photography, for example, where
culverted rail embankments only delay inundation.
Experience gained in other flooding studies, such as the
Teifi, may later be built into the computer model, or into
subjective adjustment of the results in certain instances.

Detailed description of the study area and the results
so far is unnecessary, since Figure 6 summarises these
adequately. However, it is worth noting that the short
reach of the Sence chosen includes several of the classic
modifications to natural floodplains in urban areas. For
example, the railways crossing the valley do so at a
variety of angles, and thus form a number of different cut-
off areas and potential ponding sites. Furthermore, this
variety is increased by virtue of the variation in bridge
type and aperture size. In fact, few of the apparent cut-
offs are likely to escape flooding, but the pattern and
timing of the inundation will vary quite considerably
according to the nature of the access and the distance that
the flood water may have to back-up.

The canal, following a valley-side contour, cuts off
definite tributary valleys. The bridges over these tributaries would seem to allow ready access for the flood
water to cut-off areas. Roads, it would appear from this
example, are less important, except where they are embanked,
as some more modern stretches tend to be. Inevitably, the
predictive model will also throw up anomalies like the
large quarry excavated to below the valley level, but
nevertheless most unlikely ever to flood in response to
streamflow rise.

In addition to predicting flood extent, the approach
taken in this study should allow a quantitative assessment
of lost flood storage, which may in turn be used to predict adjustments to flood stage. Furthermore, a sequence
of inundation may be hypothesised for a short reach by
taking a number of levels lower than the maximum (Figure 6
includes two levels set at an arbitrary 1.0 m and 0.5 m
below the '25 year' flood plane).

DISCUSSION

Qualitatively, the main effects of floodplain modification, even in areas where planned engineering works do not occur, would appear to be:

(a)   reduction in flood stage following enlargement, cleaning and straightening of the channel, or a decrease in flood frequency arising from the construction of embankments and the elimination of natural bank breaches.

(b)   modification to effective floodplain storage following land drainage operations, infilling and floodplain compartmentalization. These may have complex effects which may not all work in the same direction. Thus embankments across valleys may lead to an effective increase in storage upstream, whilst embankments which isolate parts of floodplains from inundation may lead to a decrease.

(c)   resultant hydrograph modifications. These may involve more nearly translatory flood wave movement in so far as this is affected by an acceleration of overbank or in-channel flow and a decrease in storage, or to the opposite effect should floodplain modifications increase effective storage.

Although it is possible to illustrate such processes and to identify the possible decrease in inundation area in particular cases, the detailed hydrological effects are less certain. For example, structures are seldom completely impermeable, and inundation may occur beyond barriers like roads and railways via ditches and gateways. Modified flood processes are locally complex, and this creates problems in theoretical flood mapping from aerial photographs. In the case of the Sence, the flood heights for the '25 year' flood themselves reflect the effects of confining floodplain structures, so that the former extent of floods cannot simply be derived by projecting flood levels ignoring such structures, because in their absence flood stage itself would differ.

Field data which allow quantitative comparison of hydrograph and inundation characteristics before and after floodplain modification appear to be unavailable, but it does seem possible that such data could be obtained. The suggestion in this paper may provide an additional explanation for hearsay observations that floodplain modifications have led to alterations in flood characteristics. This is human impact on hydrology additional to the effects of urbanization, land use and management modifications which can be explored more fully.

THE URBAN ENVIRONMENT

11

URBAN INFLUENCES ON PRECIPITATION IN LONDON

B. Atkinson

Department of Geography, Queen Mary College, London

ABSTRACT

*Man's influences on water in the air are rather speculative
and sparsely reported. Mean annual precipitation data
present a confusing picture. In North America several
cities appear to show a clear urban maximum of precipitation in agreement with earlier central European results,
whereas in Britain no such maximum has been found. More
detailed climatological analysis concentrating on convective precipitation, have revealed urban maxima, not only
in North America but also in Britain. The relative warmth
of towns, the (occasional?) higher absolute humidities in
towns, the greater concentration of pollution in towns and
the mechanical effects of towns could all contribute to
this phenomenon. The effects of greater warmth and moisture in the city are illustrated by three storms. In the
first case, the storm grew over central London and drifted
very slowly north-westward. In the second case the storm
approached London from the west and intensified as it
crossed the city. In the third spectacular case (the
Hampstead storm) the storm appeared to stagnate over
northern London. In all three cases the warmer city air
played an important role in the development of cloud
growth and consequently influenced precipitation distribution. The mechanical effects of the city are illustrated
by a storm where frictional and obstacle effects of the
buildings on the airflow induced vertical motion which
favoured precipitation over the city. Difficulties in data
collection prevent the thorough analysis of the effects of
pollution on precipitation in urban areas. American
results from Buffalo, St. Louis and Seattle all suggest
that the pollution pall is a significant source of potential condensation and ice nuclei.*

INTRODUCTION

The common interest of hydrologists and meteorologists in precipitation may be neatly partitioned: water on and in the ground concerned the hydrologist; water in the atmosphere concerns the meteorologist. Man's influences on the former are many and comparatively well-documented; his influences on the latter are rather speculative and sparsely reported. This paper concentrates on the influence of urban areas - London in particular - on precipitation amounts.

The problem may readily be formulated in terms of two questions. First, do urban areas receive more or less precipitation than the surrounding flat, rural areas? Second, if there are real differences between precipitation amounts received in urban and rural areas, what are the mechanisms causing these differences? These questions have intrigued urban climatologists for half a century but it is only within the last decade that substantial evidence for urban effects has been produced - particularly by the Metropolitan Meteorological Experiment (METROMEX) (Changnon et al., 1971). Prior to this massive observational programme, the bulk of the evidence lay in the German literature, well summarized by Landsberg (1956).

Attempts to answer the two questions may be divided into two types: empirical and theoretical. Of the two, by far the more popular approach has been the former. The empirical approach itself usually comprises two methods of analysis, either the classical climatological method or the case study. Both methods have been frequently used in the last five decades, the climatological method being the more popular. In this method, mean daily, monthly or annual values of precipitation in and around the urban area in question are mapped. If a maximum of precipitation is found over or down-wind of the urban area, it is frequently attributed to 'urban effects' without any analysis of their physical feasibility. Closely allied to this approach is the examination of long-term precipitation records both within and without an urban area. Should different trends exist between urban and non-urban areas, as found by Barrett (1964) and Changnon (1961), then again, these differences are attributed to an effect, begging the question of how this 'effect' actually manifests itself.

In contrast to the climatological approach, case studies concentrate upon the processes operating in a situation where more (rather than less) precipitation has clearly fallen over urban than rural areas. Such analyses have frequently revealed that urban areas do have some effect upon precipitation amounts. This is not to say, of course, that towns affect all types of precipitation but such studies do strongly suggest that the 'effect' is physically possible. The very few theoretical studies of the problem which exist support this viewpoint. For example, Och's (1975) model of cloud growth over St. Louis showed for the first time that the urban heat island was the primary cause of cloud initiation over the city.

In contrast to the major American investigations of urban effects on precipitation, British analyses of the

problem remain few and far between. This paper reviews
the empirical investigations of London's effects upon
precipitation.

DATA AND METHODS

London offers three major advantages to the urban climat-
ologist and particularly to the analyst of precipitation.
First, the built-up area covers over 2000 km^2, with a
diameter of about 30 km. *A priori* one would expect the
possibility of an urban effect to be increased in some
proportion to the residence time of air over the urban
area. Second, Chandler's (1965) *Climate of London*
provides a firm bed rock for any analysis of precipitation,
and finally even with the aid of data only routinely
collected by the Meteorological Office, the picture of
precipitation within London and southeast England may be
painted in some detail. Throughout the period of study
(1951-60), over 600 daily raingauges in south-east England
had a complete record (Figure 1). In addition, over one
hundred autographic raingauges in and around London
operated for some time during that decade. Any explanation
of precipitation distributions must include consideration
of the other meteorological elements such as surface and
upper air temperatures, humidities, and wind speed and
direction. Once more, south-east England provided a dense
routine network of surface stations which record these
elements variously, once, twice, four, eight or twenty-
four times per day (Figure 2). Upper air conditions were
available from Crawley, some 40 km south of London.
     For reasons outlined below, the analysis was confined
to thunder rainfall and data on thunderstorm occurrence
were kindly made available by the Electrical Research
Association in Leatherhead, Surrey. The volunteer network
(Figure 3) was quite impressive but it is important to note
that many records covered less than the full decade.
     In addition to the above data, both 'sferics' and radar
were used to locate thunderstorms. The former has limited
use due to inaccuracies in the 'fix' but the latter proved
most valuable in locating the storms, despite rather
primitive equipment and far from continuous operation.
     In the light of the limitations of previous studies, it
was decided to restrict the investigation to convective
precipitation. It was thus possible to concentrate on the
ways in which urban areas may affect the already quite
well-known mechanisms of one type of precipitation, without
the hindrance of other types. In addition, at their
simplest, convective clouds comprise thermals, usually from
the ground, and it is well known that towns are frequently
warmer than their surrounding rural areas. Therefore it is
not unreasonable to suggest that towns may be a preferred
source area for convective clouds. Unfortunately, there is
no easy way to observe all convective clouds in detail on a
routine basis, so for practical purposes attention was
restricted to thunderstorms i.e. convective clouds which
give lightning and thunder. Both the Climatological
Observers of the Meteorological Office and the Electrical

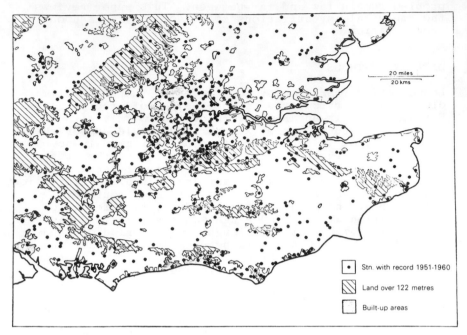

Figure 1.   British Rainfall stations in south-east England,
            1951-66.

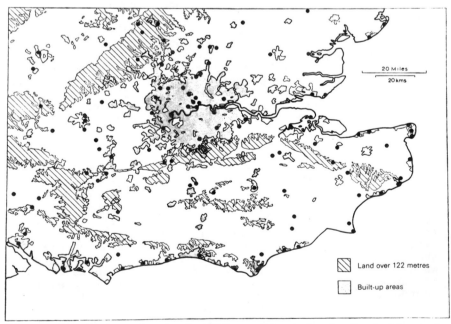

Figure 2.   Meteorological Office climatological stations in
            south-east England, 1951-60.

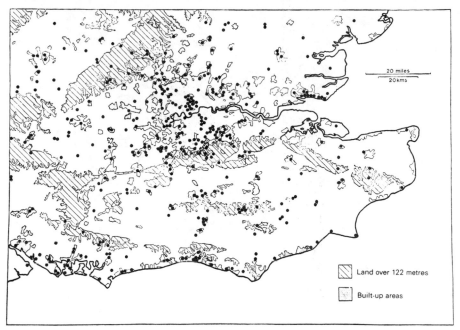

Figure 3.   Electrical Research Association thunderstorm
       observers in south-east England, 1951-60.

Research Association observers provided quite detailed
information on the location, duration and intensity of
thunderclouds over south-east England during the period
1951-60.
    With the data outlined above it proved possible to
investigate the problem by both climatological and case-
study methods.   The results and implications of these
analyses are reviewed below.

RESULTS

Climatological investigation (Atkinson, 1968, 1969)
revealed maxima of both days with thunder overhead and
total thunder rainfall in east central London (Figure 4).
Within the central city area over 1000 mm of thunder rain
were received in the whole decade in comparison with about
700 mm in surrounding rural areas of similar altitude.
The bulk (70-90%) of these amounts was received in summer
(May to September inclusive).   The relative magnitude of
the urban maximum varied with precipitation intensity
(more specifically the amount of precipitation associated
with each thunder outbreak).   In 'light' (0.01 to 12 mm
per outbreak) intensities, the London area had at least as
many outbreaks and as much precipitation as the upland
areas of the Chiltern Hills and parts of the North Downs.
The urban area showed the largest increases in outbreak
frequency and precipitation amount for intensities of

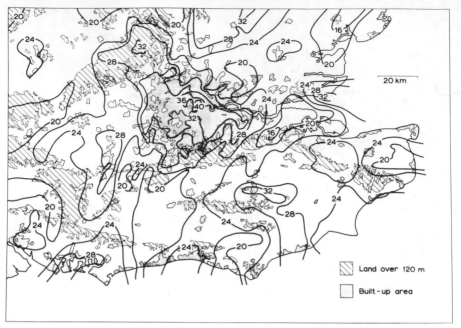

Figure 4.   Total thunder precipitation (inches) in south-
east England, 1951-60.   Approximate metric equivalents
are: 20 in - 500 mm;   24 in - 600 mm;   40 in - 1000 mm.

12-65 mm per outbreak.

A preliminary attempt to investigate the mechanisms
responsible for the urban maximum involved the classification of the thunderstorms by their synoptic origins
(Atkinson, 1966, 1967).   This soon revealed that the urban
maximum appeared most strongly within six types of synoptic
situation, viz: northerly, north-westerly, westerly, and
south-westerly airstreams, the so-called 'slack' circul-
ation and warm fronts.   In all six cases, the thermodynamic
structure of the atmosphere was particularly favourable to
the growth of cumulonimbus clouds over a warm, rough
surface such as presented by London's urban area.   Potentially profitable case studies became readily available
from this synoptic analysis.

Four factors could possibly account for the greater
precipitation amounts over London.   They have long been
recognised as higher urban temperatures - the heat island,
higher urban humidities, higher concentrations of condensation and ice nuclei in the polluted urban air and the
mechanical effect of the urban surface.   All four factors,
but particularly the first, readily apply to convective
precipitation.   It is well-known that a vertically unstable
atmosphere favours convection and, if sufficient moisture
be present, cumulus growth.   Instability itself is related
to the vertical temperature gradient and any low-level heat
source will encourage the development of a steep lapse rate
and thus instability.   Consequently, the well-documented

urban heat island would be expected to help rather than
hinder convection. Chandler (1965) has shown that London
may have a heat-island both day and night and Behilak
(1967) has measured super-adiabatic lapse rates in the
lower layers of the urban atmosphere. More explicit confirmation of the effects of the heat island lies in the
METROMEX observations of radar echoes over the major
industrial area of St. Louis (Changnon, 1971).

Dry convection cannot of course produce precipitation.
The formation of cloud and precipitation occurs only if
the thermals are originally quite moist and in turn their
humidity content is, to a large degree, a function of the
rate of evaporation from the underlying surface. Until
comparatively recently, and in the absence of good
measurements, urban air was considered to be drier than
rural air, the reasoning being that run-off would be
higher in urban areas due to the impervious surface and so
evaporation rates must be less. In the last decade we have
learnt (Chandler, 1967; Bornstein et al., 1972; Kopec,
1973) that absolute humidities may in fact be 2-3 mb
higher than in the surrounding rural areas. Relative
humidities are still frequently lower in the city due to
its higher temperatures.

Data on condensation and ice nuclei in urban air are
scarce. In America Kockmond and Mack (1972) and Semonin
and Changnon (1974) have measured increases in condensation
nuclei as air crossed the cities of Buffalo and St. Louis
respectively. Similarly, Braham and Spyers-Duran (1974),
despite observations apparently to the contrary, consider
that the number of ice nuclei in air also increases as it
crosses the city. Some doubt exists here due to a possible
fault in the relevant instrument. The importance of all
these measurements lies in the effects that the increases
in nuclei have upon the spectrum of cloud droplet sizes.
Opinions differ on this matter. As early as 1957 Gunn and
Phillips (1957) surmised from laboratory experiments that
for a given amount of water vapour, any increase in nuclei
would lead to a greater number of smaller droplets than
previously. Consequently cloud may increase, but the
smaller drops decrease the likelihood of precipitation by
coalescence from such cloud. In essence these conclusions
agree with those of Kockmond and Mack (1972) and Fitzgerald
and Spyers-Duran (1973) whose observations revealed a
narrowing of droplet spectra in clouds which passed over
urban areas. In contrast, Hobbs et al. (1970) observed a
broadening of the droplet spectra in similar situations.
In effect this means the creation of larger droplets, the
existence of which encourages the coalescence mechanism.
These contradictory findings stress the need for more
definitive measurements of 'urban' nuclei.

Under the heading of 'mechanical' effects, the urban
area acts as an obstacle, on the one hand, and causes
frictional convergence on the other. Angell et al. (1973)
found that in strong winds of about 13 m s^{-1} Oklahoma City
acted as a barrier to flow and induced upward velocities
to air parcels of up to 70 cm s^{-1} at the 400 m level over
parts of the city. The frictional convergence effect was
observed by Ackerman (1974) in St. Louis. From the

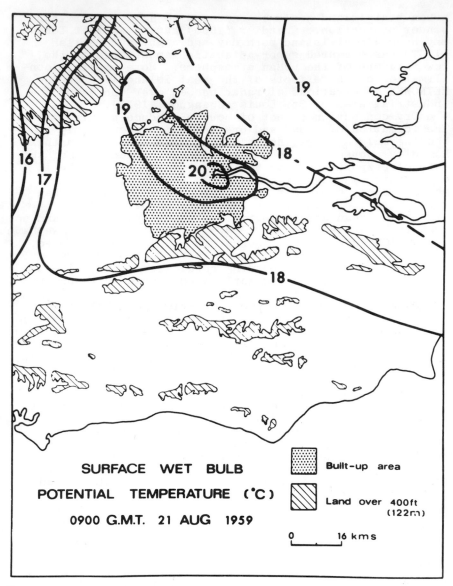

SURFACE WET BULB
POTENTIAL TEMPERATURE (°C)
0900 G.M.T. 21 AUG 1959

Built-up area

Land over 400ft
(122m)

0   16 km s

Figure 5A.   Surface wet-bulb potential temperature (°C) at
    0900 GMT   21 August 1959.

horizontal convergence Ackerman calculated upward velocities
of 2-6 cm s^{-1}.   In both cases, the vertical motions would
encourage cloud growth, either directly by taking the form
of thermals, or indirectly, by de-stabilizing the lower
atmosphere so that later thermals would rise more readily.
    The relative importance of these four factors is well
summarized in Battan's (1965, p 79) words 'the micro-
physical properties of ... clouds are not of dominant impor-

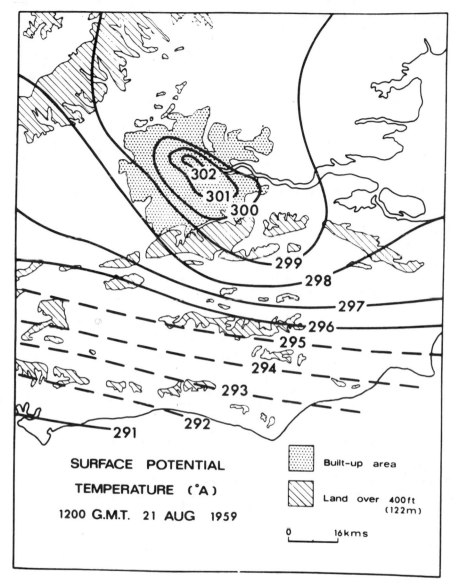

Figure 5B. Surface dry-bulb potential temperature ($^\circ$K) at
     1200 GMT 21 August 1959.

tance in determining how much precipitation reaches the
ground.' In convective clouds at least, ' those properties
of the atmosphere which govern the characteristics of the
updraughts are the ones which chiefly control the quantity
of rainfall...' (Battan, 1965, p 84). In the light of
Battan's statement (which is supported by the bulk of our
knowledge of cloud physics), the effects of urban temperatures, humidities and windfields on particular storms over

London have been analysed.

Before looking at these cases in more detail it is important to note the slightly different roles of dry-bulb (θ) and wet-bulb (θ_w) potential temperatures in the development of cumulus and cumulonimbus clouds. Substantial cloudy convection is encouraged by a strong vertical lapse of wet-bulb potential temperature and this lapse may be due to the existence of either cool, dry air at high levels or warm, moist air at low levels, or both. At any one time, the synoptic and meso-scale circulations are the prime determinants of vertical gradient of θ_w, but in the lowest 500 m of the atmosphere it is possible that effects of, for example, urban areas or forests, could make themselves felt. Their impact would, of course, be comparatively local. Once the distribution of θ_w favours convection, it is the distribution of θ which actually determines the source of thermals. For example, individual cooling towers, buildings or ploughed fields may release thermals (due to high θ values) which would rise rapidly because of the favourable θ_w distribution. As urban areas are warmer, and occasionally more humid than the surrounding rural areas, not only higher values of θ, but also higher values of θ_w are frequently observed within the city. The important effects of these distributions are examplified below.

Three storms, namely those occurring on 9 September 1955, 21 August 1959 and 14 August 1975, have been analysed in detail, to assess the effects of urban temperatures and humidities. The last two storms originated over London's urban area and gave maximum precipitation amounts of 68 and 170 mm respectively. In both cases London lay in an area of convergence which encouraged uplift, but the primary reason for convection in each case was the higher θ and θ_w over the urban area (Figure 5). On 21st August 1959 the radar clearly revealed cloud growth over east-central London, the area with the highest maximum temperatures and the area which received the heavy precipitation (Atkinson, 1970). In the famous Hampstead storm of 14 August 1975, initial cloud growth occurred over the higher ground of the Heath but secondary 'cells' developed over lowland central London due to lifting of the warm air there by the cold outflows from the original cells (Atkinson, 1977). The remarkable localization and intensity of this storm remain largely unexplained.

In contrast to the above storms, that of 9 September 1955 originated to the west of London and moved over the city (Atkinson, 1971). As the storm passed over the urban area cloud tops rose at about 6 m s^{-1} resulting in a local precipitation maximum of about 25 mm over central London. This rapid cloud growth was due to the drawing into the storm of comparatively warm, moist air ($\theta_w = 16^{\circ}$C) which lay over the urban area. In contrast, rural values of θ_w were about two degress lower.

The mechanical effects of urban areas are more difficult to analyse, primarily due to lack of the right type of data. Nevertheless, the storm on 1 September 1960 appears to have been affected by the strong convergence field (10^{-4} s^{-1}) over London's urban area. Calculations based on observed cloud behaviour indicate that clouds formed by this mechanism could in fact precipitate over the city despite quite strong ambient winds (Atkinson, 1975).

CONCLUSION

The evidence strongly suggests that London's urban area may at times be an important factor in producing convective precipitation. These conclusions are supported by both the empirical and theoretical findings of the METROMEX experiment. At present we are unable to produce any forecasting rules which would be of value to, among others, hydrologists interested in urban floods. Yet the general acceptance of the physical reality of the effect provides a further important tool for the weather forecaster, complementing his knowledge of, for example, orographic and coastal effects on precipitation.

from information on distribution

... small-scale plots (not provided over the reproductive
intensity, Aldous and Waddle (Jackson, 1965).

DISCUSSION

The evidence strongly suggests that London's ... ... may
exist a ... important factor in producing ... ...
... population. ... these predictions suppress ... law
... ... and ... ... instance of the ... ...
... the ... Alternatively, one ... to produce ... ...
... control ... which would be ... ...
... ... in ... ...
... ...
... ... important ... ...
... ...
... ...

12

THE HYDROLOGICAL IMPACT OF
BUILDING ACTIVITY: A STUDY NEAR EXETER

D.E. Walling

Department of Geography, University of Exeter

ABSTRACT

Building activity associated with urban development can give rise to marked changes in hydrological response, particularly in terms of erosion and sediment yield. Studies of construction sites in the eastern United States have documented increases in sediment yield as high as 375 fold. In order to provide information on the potential hydrological impact of building development under British conditions, a study has been undertaken in a small catchment on the margins of Exeter. Records from the period August 1968 to March 1971, when the basin was essentially undisturbed, have been used to calibrate its response and to assess the subsequent changes in storm runoff generation, suspended sediment yield and solute levels that occurred during 1973 when 25% of its area was influenced by building activity. On average, storm runoff volumes and peak flows increased 2-4 times, suspended sediment concentrations 5 fold, sediment loads 5-10 fold and specific conductance levels 30%.

INTRODUCTION

The hydrological effects of urbanization have attracted much attention in recent years from both academic and practical viewpoints and this topic has formed an important component of international research programmes initiated under the I.H.D. (e.g. UNESCO, 1974). It is now clearly apparent that the urban landscape exhibits a completely different hydrological response to that of rural areas (e.g. Cohen et al., 1968; Leopold, 1968). Relatively less attention has been given to the hydrological impact of the actual process of change from rural to urban or suburban conditions and of building activity in particular. In the United States approximately 0.04 per cent of the land area

(c. 2890 km^2) is urbanized each year (ASCE, 1975) and the
adverse environmental effects have attracted the attention
of several researchers. Probably the greatest hydrological
change conditioned by construction activity is in sediment
yield. Guy (1965), working in Washington D.C., found that
an additional 10 tons of sediment would be transported by
streams in the area for every person added to the city.
Subsequent work (e.g. Guy, 1970; Guy, 1972; Wolman and
Schick, 1967; Yorke and Davis, 1972) has added further
evidence of the extent of increases in sediment yeild and
the undesirable effects of construction activity. The
eroded sediment can give rise to serious problems in terms
of downstream desposition and increased stream turbidity
can cause serious degradation of water quality with
associated ecological and economic consequences. In view
of the inclusion of sediment in pollution control legislation, serious consideration is now being given to the
problem of control of construction site erosion in many
areas of the United States (e.g. ASCE, 1975; Brandt, 1972).

Once urban conditions have been established, sediment
loads will, in most cases, decline rapidly to a level
below that characterising neighbouring rural areas. The
period of building construction will therefore stand out
markedly in terms of its effects on sediment yield.
Building activity can also be expected to cause changes in
storm runoff dynamics, since the infiltration capacity of
the disturbed areas will generally be considerably less
than those of natural surfaces. However, the imperviousness will normally increase with the establishment of the
urban area and the modification of runoff processes will
be maintained and intensified. Few attempts have been
made to quantify the effects of building activity on stream
levels because it does not normally constitute a serious
problem. In some cases point pollution sources associated
with the construction work will cause increases in solute
transport, but disturbance of soil and regolith and disruption of the nutrient cycle by vegetation destruction
could also be expected to cause increased solute loadings.

Some of the increases in sediment loads associated
with building activity documented in the eastern United
States are summarised in Table 1. Increases in annual
suspended sediment yields of over 300 fold have been found.
Little attention has been paid to the impact of building
activity on erosion and sediment yield and other hydrological processes in Britain, possibly because of the
assumption that the scale and nature of the urbanization
process and the character of the terrain and the hydrometeorological conditions are not conducive to the marked
changes and problems that occur in other areas. In an
attempt to document the magnitude of such hydrological
modifications in an area of Britain, a study was initiated
within a zone of suburban development on the north eastern
margins of Exeter (Walling and Gregory, 1970; Walling,
1974).

The investigation involved the use of an instrumented
catchment and the single watershed approach was adopted.
Measurements of streamflow, sediment discharge and solute
yield were made whilst the catchment was in an essentially

Table 1. Increases in catchment sediment yield consequent upon building activity documented by several studies in the metropolitan Washington and Baltimore region of the eastern United States. (Based partly on Chen, 1974)

Location	Catchment Area (km²)	Ad/D[1]	Sediment yield (Tonnes.ha⁻¹.yr⁻¹)	Approximate increase[2]	
				Catchment	Disturbed Area
Lake Barcroft, Faifax, Va Holeman and Geiger (1965)	37.6	0.07	5.7	x2.8	x30
Scott Run, Fairfax, Va. Vice et al. (1969)	11.8	0.07	10.0	x5	x51
Manor Run, Norbeck, Va. Yorke and Davis (1972)	2.6	0.11	10.6	x5.3	x40
Kensington Md. Guy (1965)	0.24	0.35	32.5	x16.3	x45
Oregon Branch, Cokeysville, Md. Wolman and Schick (1967)	0.61	1.0	252	x126	x126
Minebank Run, Townson, Md. Wolman and Schick (1967)	0.08	1.0	281.3	x140	x140
John Hopkins University Wolman and Schick (1967)	0.0065	1.0	487.7	x375	x375

1. Ad/A represents the proportion of the catchment area being developed.
2. These values have been calculated assuming an average sediment yield for undisturbed catchments in this area of 2 tonnes.ha⁻¹.yr⁻¹ The magnitude of the catchment increase reflects the Ad/A value for the basin. No account has been taken of sediment delivery considerations in calculating disturbed area increase – the increase necessary to account for the yield from the overall basin.

undisturbed state, in order to calibrate the natural hydrological response. Subsequent departure from this natural
response was assessed using this calibration.

THE STUDY CATCHMENT

The study basin (Figure 1A), often referred to as the
Rosebarn catchment, extends over an area of 0.26 km^2 and
has an absolute relief of 83 m. It is underlain by shales
of the Carboniferous Culm Measures which weather on the
surface to form a deep heavy clay. In its undisturbed
state the perennial channel network comprised approximately
950 m of well-defined channel. This was characteristically
incised into the clay, so that the banks were formed of
cohesive clay material, whereas the channel floor was in
places composed of angular fragments derived from the underlying bedrock. When the catchment was originally instrumented in the summer of 1968, about 8.7 per cent of its area
was developed for residential purposes, but because this
development was situated on the divides it was judged to
exert a limited effect on the sediment and solute response
of the basin. Vegetation and land use in the remainder of
the catchment consisted primarily of permanent pasture.

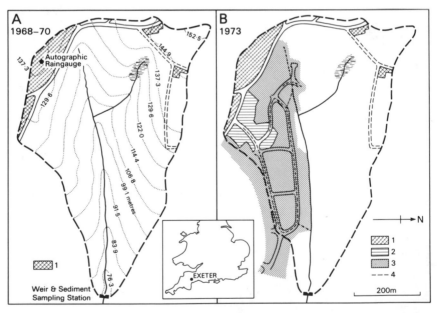

Figure 1. The study catchment showing the original built-
up area (1), the area of housing completed by 1973 (2),
the area of building activity during 1973 (3), and the
network of storm water sewers (4).

Precipitation input was monitored with a daily auto-graphic gauge situated within the boundary of the catchment (Figure 1). Streamflow at the basin outlet was gauged by a compound sharp-crested weir in association with a water level recorder equipped with a daily chart. The relatively impermeable clay surface generates a 'flashy' runoff response, with individual rainfall events producing clearly defined streamflow rises. Sampling of sediment and solute yield was carried out at a point immediately upstream of the pool associated with the measuring weir. Initial observations during flood events suggested that bedload transport in the channel was minimal and this was confirmed by the lack of an accumulation of coarse debris behind the gauging structure. Attention was therefore focused on suspended sediment transport and detailed measurements were undertaken. Initially, a US DH48 sampler was used for sampling on both a systematic time basis and during storm events, but the need for comprehensive data on the sediment loads associated with individual storms stimulated the development and construction of an automatic pumping sampler for use during storm events (Walling and Teed, 1971). A sampler which operated automatically at 10 minute intervals during storm events was installed near the catchment outlet in early 1970. Solute sampling has been carried out in association with suspended sediment sampling and the bottles provided by the pumping apparatus have also been used for this purpose.

Laboratory determination of suspended sediment concentration has been based on vacuum filtration through glass fibre (Whatman G.F.C.) filter circles. Analysis of solute samples has been primarily restricted to determination of specific electrical conductance although analysis of individual ions has also been undertaken occasionally.

The essentially undisturbed condition of the catchment persisted until March 1971 when sporadic building activity commenced. The state of the catchment during 1973 is shown in Figure 1B. At that time, approximately 25 per cent of its area was disturbed by construction activity. A network of roughly surfaced service roads had been installed and areas of bare disturbed soil were associated with both the road sides and with site preparation for houses. The drainage network had also been significantly modified by the installation of a system of storm water sewers for collecting road and pavement drainage (Figure 1B).

RESULTS

The records available for the period August 1968 to March 1971, when the catchment was in an essentially undisturbed condition, provide a means of calibrating its natural behaviour. Many problems are associated with calibration procedures based on a short period of record, but it is thought that the approaches adopted afford a worthwhile means of establishing baselines against which subsequent changes can be evaluated. In this paper the departures from natural response associated with 1973, when approximately 25 per cent of the catchment area was disturbed by construction activity, will be considered.

Fundamental in reflecting and conditioning changes in hydrological processes within a catchment are the storm runoff dynamics. A detailed appraisal of the various changes occurring during earlier phases of development in the basin has been published elsewhere (Gregory, 1974) and in this discussion attention is primarily focussed on storm runoff volumes and peak flow rates associated with individual storm hydrographs.

The storm hydrograph separation procedure advocated by Hibbert and Cunningham (1967) was used to calculate the quickflow or storm runoff volumes and peak storm runoff rates for all simple hydrographs that occurred during the calibration period. Stepwise multiple regression was subsequently employed to derive prediction equations relating these two measures of storm runoff to various hydrometeorological variables. When deriving a prediction equation of this type for calibration purposes, it is essential that the independent variables should themselves not change as a result of subsequent modifications of hydrological conditions. Measures relating to the flow levels preceding individual storm events, for example, do not satisfy this criterion.

Table 2. Stepwise multiple regression equations derived to predict storm runoff response under undisturbed conditions.

Dependent variable		Equation
Log Peak Quickflow	=	$-1.2384 + 2.0198$ log Rain -0.6662 log Duration -0.0067 SMD $+0.7597$ log API, $R = 0.92$
Log Storm Runoff	=	$-0.3367 + 2.2235$ log Rain -0.4849 log Duration -0.0071 SMD $+0.7859$ log API $+0.1734$ Cosday, $R = 0.97$

$n = 115$

Where: Rain = storm rainfall (mm); Duration = storm duration (h); SMD = soil moisture deficit (mm); API = antecedent precipitation index (mm); Cosday = cos (radians) 2π day/365.

The prediction equations are listed in Table 2. Logarithmic transformations have been applied to many of the variables, to meet data normality requirements. Inevitably, problems of interdependence amongst the independent variables included in the equations exist. However, these are not serious, since the equations are being used primarily for prediction rather than inference of process mechanisms. Nevertheless, the resultant equations conform to a physically-based system, because runoff peaks and volumes are positively associated with rainfall amount and the A.P.I. index and inversely related to storm duration and S.M.D. The Cosday variable included in the second equation represents an attempt to derive a simple seasonal index, since this will exhibit a maximum value (+1) at the beginning or end of the year and a minimum value (-1) at the end of June. The positive coefficient found here indicates a tendency for maximum storm runoff volumes to occur during the winter months.

The prediction equations presented in Table 2 have been applied to 121 storm events that occurred during 1973, in order to estimate the natural response had building activity not taken place. Care has been exercised to ensure that the equations have not been used for independent variable values outside the bounds of those used to derive them. The range of increases in peak quickflow discharge and quickflow volume are portrayed in Figure 2.

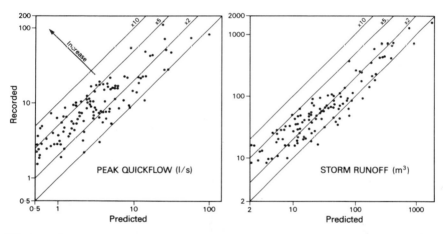

Figure 2. Comparison of the storm runoff response during 1973 with estimates of the equivalent natural response.

Increases of over 8 fold are indicated, although the average increases are 2.4 and 3.5 times for storm runoff amount and peak flow rate respectively. Inspection of Figure 2 does not reveal any clear tendency for maximum increases to be associated with either low or high magnitude events.

A change in hydrograph shape, notably an increase in peakedness is indicated by greater increases in peak flow rate than in runoff volume. This is not unexpected in view of the modification of the drainage network by construction of storm water sewers. To explore this trend further, unit hydrographs have been derived both for the calibration period and for 1973. A graphical technique has been used for this purpose and the resultant hydrographs depicted in Figure 3 represent the hydrographs of direct runoff resulting from 10 mm of effective rainfall

generated over the basin at a uniform rate during the
specified period. Many criticisms can be made of unit
hydrograph theory, particularly in terms of the assumption
of linearity of the storm runoff process (e.g. Chow, 1964),
but their use in this context is thought to be justified.
Comparison of the unit hydrographs for the calibration
period and for 1973 indicated an increase in peak flow
rates of almost 4 times for short duration storms, with
this increase diminishing for longer durations. It must
be emphasised that these increases relate solely to changes
in the delivery and routing characteristics of the catchment, because equal runoff volumes are necessarily
assumed for the two periods.

Change in time of rise are also associated with the
increases in hydrograph peakedness demonstrated in Figure
3. Values are almost halved for the short duration storm
events and they represent an extremely flashy response.
An increase in the speed and efficiency of removal of
surface runoff could also be expected to condition changes
in the recession characteristics of storm hydrographs and
reductions in hydrograph base-time are clearly evident in
Figure 3.

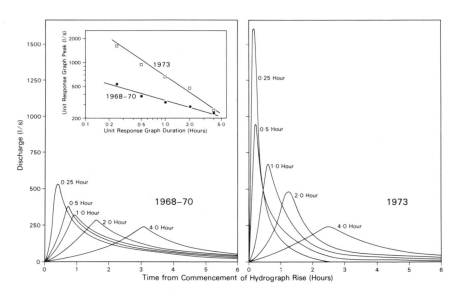

Figure 3. Unit hydrographs for the calibration period and
for 1973.

The existence of disturbed areas and changes in storm
runoff dynamics could be expected to condition increases
in sediment yield from the catchment. Two major approaches
to assessing these changes have been adopted: first.
consideration of suspended sediment concentration/discharge
relationships or ratings for the catchment and secondly,
evaluation of the sediment yield characteristics of

individual storm runoff events.

The suspended sediment concentration rating has been used in many studies to characterise suspended sediment production within a drainage basin (e.g. Bauer and Tille, 1967; Walling, 1971), and the rating plot for this basin during the calibration period is depicted in Figure 4A.

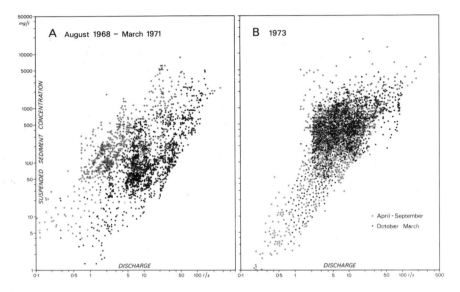

Figure 4. Suspended sediment rating plots for the calibration period (A) and for 1973 (B).

The scatter shown by this plot largely precludes any attempt to represent the rating by a single straight line, the proportion of the total variance explained by such a simple relationship being only 33 per cent (Table 3).

Table 3. Suspended sediment concentration/discharge relationships for the study catchment.

Data Set	Calibration period		1973	
Total	$Conc = 40.18 \ Q^{0.599}$	$r = 0.58$	$Conc = 50.82 \ Q^{0.908}$	$r = 0.66$
Summer	$Conc = 59.73 \ Q^{0.816}$	$r = 0.75$	$Conc = 42.20 \ Q^{1.025}$	$r = 0.68$
Winter	$Conc = 14.30 \ Q^{0.810}$	$r = 0.71$	$Conc = 85.70 \ Q^{0.664}$	$r = 0.58$

A small proportion of the scatter can be explained in terms of a seasonal effect, because zoning of samples taken during summer (April – September) and winter (October – March) is apparent. Separate regression lines have been fitted to the data for the two seasons (Table 3). The

higher concentrations in summer can be tentatively
explained by the increased incidence of intense convec-
tional rainfall, the occurrence of dry surface conditions
and the reduced baseflow component of individual storm
discharge values during that period.

Despite the limitations of the rating plot and associated regression lines for providing a precise character-
isation of the response of the catchment under natural
conditions, an attempt has been made to compare the rating
data for the calibration period with that for 1973 (Figure
4B). An obvious contrast between the two plots is the
increase in the magnitude of the maximum recorded concentrations during 1973 and a general increase in concentration levels during this period of between 2 to 5 times.
Also important is the lack of differentiation between
summer and winter samples exhibited by the 1973 data.
This cannot be related to varying degrees of surface disturbance during the year or to changes in rainfall character between the calibration period and 1973. It can,
perhaps, be provisionally interpreted in terms of a reduction in the seasonal differentiation of erodibility
occasioned by the occurrence of large areas of bare soil.

An attempt to assess the general level of increase in
sediment concentrations between the two periods has been
made in Figure 5 by comparing the general and seasonal
straight line rating relationships for the two periods
detailed in Table 3. Figure 5A, which presents the rating
lines for the complete data sets, demonstrates a general
increase in concentration for a given discharge of 2 times
at medium flows rising to 5 times at high flows. In
Figure 5B the seasonal rating lines are shown and a general
increase of winter concentrations by 3 times is apparent.

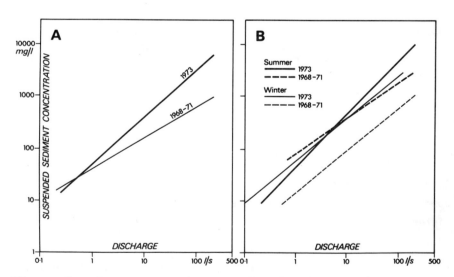

Figure 5. Comparison of the general (A) and seasonal (B)
straight line rating relationships for the calibration
period and for 1973.

Increases for the summer data set are less marked, being generally less than 2 times. The greater increases evident for the winter season are the result of the disappearance of the distinction between summer and winter samples noted previously and which has been superimposed on a general overall increase.

An important problem associated with the use of rating plots and equations for assessing increases in sediment yield caused by changes in catchment condition is that the runoff regime will in most cases also alter. Increased storm hydrograph peaks and storm runoff volumes during 1973 have already been described. The above estimates of increase in sediment yield based on direct superimposition and comparison of rating plots, are therefore probably conservative in magnitude.

A more meaningful indication, both of the pattern of sediment yield from a drainage basin and the changes consequent upon surface disturbance, can be provided by consideration of the concentrations and yields associated with individual storm events. By comparing the suspended sediment yields produced by particular rainfall inputs, both before and during urbanization, changes may be evaluated with respect to a control (i.e. rainfall) which has itself not changed. The data available for the calibration period, particularly those provided by the pumping sampler, have been used to evaluate the sediment yield

Table 4. Multiple regression equations derived to predict storm period sediment yields under disturbed conditions.

Dependent variable		Equation
Log maximum concentration	=	$1.2046 + 0.8144 \log KI + 0.3212$ Sinday, $R = 0.85$
Log mean concentration	=	$1.1772 + 0.6176 \log KI + 0.2500$ Sinday $- 0.2064$ Cosday $- 0.2195$ log Temp, $R = 0.82$
Log total load (kg)	=	$- 3.5245 + 2.9447 \log K - 0.0108$ SMD $- 0.4126 \log$ Duration $+ 0.5034 \log$ API, $R = 0.93$
Log maximum load (g.s^{-1})		$- 2.1831 + 1.7668 \log KI + 0.3254$ Sinday $- 0.0089$ SMD, $R = 0.92$

$n = 102$
Where: K = Kinetic energy of storm (Joules. cm^{-1}.m^{-2}); KI = K x maximum 30 min rainfall (mm); Temp = soil temperature (^{0}C); Duration = storm duration (h); SMD = soil moisture deficit (mm); API = antecedent precipitation index (mm); Sinday and Cosday = sin or cos (radians) 2πday/365.

characteristics of individual storm events. For each
event which produced a single peaked hydrograph, values of
peak and mean concentration, total load and maximum instantaneous load associated with the storm runoff component
have been calculated using a computer routine. In all,
data from 102 events of varying magnitude were processed.

A similar approach to that used for storm runoff has
been employed to derive prediction equations relating the
four indices of sediment yield to various measures of
rainfall character, catchment condition and season. These
would be applicable to the catchment in its natural state
and would enable changes consequent upon subsequent surface
disturbance to be evaluated. The final regression
equations are listed in Table 4. For each sediment yield
parameter, the most important control was a measure of the
rainfall energy; either an estimate of the kinetic energy
based on the data produced by Wischmeier and Smith (1958)
or the product of this and maximum 30 minute rainfall

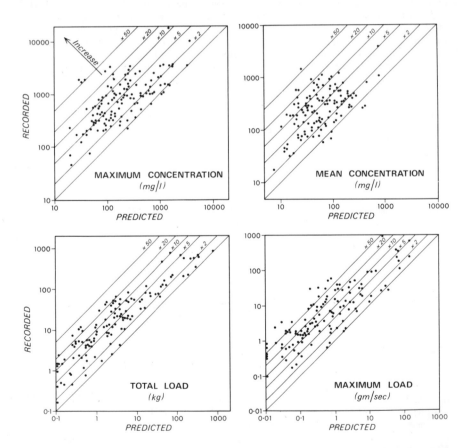

Figure 6. Comparison of the sediment yield characteristics
of storm events during 1973 with estimates of the
equivalent natural response.

intensity, a parameter used with considerable success in
the Universal Soil Loss Equation (Wischmeier and Smith,
1965). The other variables incorporated in the multiple
regression equations represent measures of moisture status
and surface condition (API and SMD) and general indices of
season (sinday and cosday).

The degree of explanation provided by these prediction
equations varies considerably (Table 4), but they have
been used as the best means available at present for
estimating what the sediment yield characteristics of
storms which occurred during 1973 would have been, had
building activity not occurred. Again, particular care
has been taken not to apply the equations outside the
range of the original independent variables. A comparison
of the observed sediment yields of 121 storms which
occurred during 1973 with the values predicted using these
equations is provided in Figure 6. Mean increases were
5.7, 6.5, 8.5 and 21.3 times for maximum and mean concentrations, total loads and maximum instantaneous loads
respectively. Values of load exhibit greater increases
than those of concentration, because they reflect increases
in both concentration and storm discharge. Maximum
instantaneous loads show the greatest increase, with values
of over 50 fold occurring on occasions.

Inspection of the data presented in Figure 6 does not
reveal any marked tendency for preferential increases in
concentrations and loads of a certain magnitude, although
it could be suggested that increases are slightly greater
in the case of events of low and medium magnitude. This
impression is supported by a multiple regression analysis

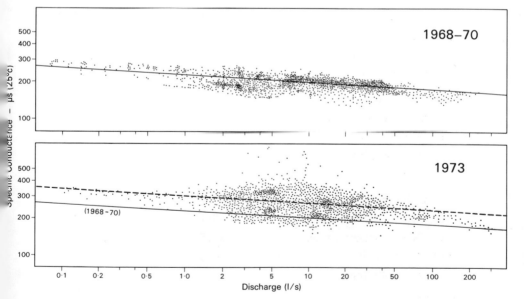

Figure 7. Comparison of the specific conductance/stream-
flow relationships for the calibration period and
for 1973.

of increase rates in relation to the various independent variables listed in Table 4. Increases in concentration and maximum load were inversely related to rainfall amount. However, there was also a tendency for maximum increases in concentration to be associated with storms of high kinetic energy, conditions of low moisture status, and the winter months. The previous comparison of rating data also indicated a greater degree of increase in sediment concentrations during the winter months.

Catchment disturbance consequent upon building development could also be expected to condition changes in the solute content of streamflow. Attention has been focused on the total dissolved solids concentration of the water, as indexed by measurements of specific conductance. Inspection of the data collected during the calibration period demonstrated that the catchment exhibited the commonly-found inverse relationship of solute concentration and streamflow, with levels falling during storm runoff events. The specific conductance/discharge relationship has therefore been used to characterise the natural solute response of the study catchment (Figure 7). During calibration, specific conductance values ranged from $300\mu s.cm^{-1}$ at low flow to $135\mu s.cm^{-1}$ during peak stormflow.

Table 5. Estimated average changes in hydrologic response of the Rosebarn catchment due to building construction

Basis for comparison	Increase	
	Catchment	Disturbed area
STORM RUNOFF		
Storm runoff peaks	3.5	11.0
Storm runoff volumes	2.4	6.6
SEDIMENT YIELD		
Sediment rating data		
(a) Total plot	x2-5	x5-17
(b) Summer plot	< x2	< x5
(c) Winter plot	x3	x9
Storm period data		
(a) Max. concentration	x5.7	x20
(b) Mean concentration	x6.5	x23
(c) Total load	x8.5	x31
(d) Maximum load	x21.3	x82
SOLUTE YIELD		
Specific conductance	av.30%	av.120%

Figure 8. Building activity on the northern outskirts of Exeter during 1976. The study catchment occupies the right hand background of the photograph.

A similar conductivity/discharge relationship has been constructed using the solute measurements taken during 1973 (Figure 7). A general inverse trend is again found, although the plot exhibits considerably more scatter. Some storm hydrographs of intermediate magnitude increases in specific conductance and this accounts for the scatter of high conductivity values associated with medium discharge levels. Values of up to $750\mu s.cm^{-1}$ were recorded. Comparison of the least-squares fitted straight line relationships for the two periods indicates a general increase in conductivity levels of 40% at low flows decreasing to about 25% at medium and high flows. This can be attributed to disturbance of the soil and regolith, since the introduction of extraneous material as solute sources is thought to have been minimal.

CONCLUSIONS AND IMPLICATIONS

The results provided by the various comparisons of catchment response during the calibration period with that of 1973 are summarised in Table 5. As a result of building activity in part of this drainage basin, average increases of 2-4 fold for storm runoff peaks and volumes, 5-10 fold for suspended sediment loads, 5 fold for sediment concentration and 30% for specific conductance levels have been monitored. Because these increases are based on a catchment with only 25 per cent of its area disturbed by building activity, it is reasonable to argue that increases within the disturbed area itself would be much greater (Table 5). The response of the natural portion of the catchment would 'dilute' the extremes associated with the area of building activity.

The changes in hydrological response documented for this small drainage basin are not as spectacular as some of those reported from the United States (Table 1) where building development can involve extensive areas of surface disturbance. Nevertheless, they should be borne in mind as a potential problem, particularly where large-scale development occurs and where the proportion of basin area which remains undisturbed, and therefore serves to dilute the increases produced by the construction areas, is small. The extension of urban areas in England and Wales predicted for the period up until the year 2000 may involve about 4 per cent of the surface area. The appearance of building activity in the landscape (e.g. Figure 8) should evoke questions as to the potential impact on downstream areas of increased flood flows, increased sediment loads and concentration and changing solute levels.

ACKNOWLEDGEMENTS

The Rosebarn study was initiated by the author and Professor K.J. Gregory (now of Southampton University).

Thanks are due to him for his enthusiastic co-operation and to the Natural Environment Research Council who supported part of the investigation with a Research Grant.

13

THE EFFECT OF URBANIZATION
ON FLOOD MAGNITUDE AND FREQUENCY

J.C. Packman

Institute of Hydrology, Wallingford, Oxfordshire

ABSTRACT

The problems of estimating the degree of modification of flood response owing to urban development are discussed, namely how the increase in percentage runoff, the reduction in lag time, and the flattening of the flood-frequency growth curve are related to the increase in impervious area, the improvements made to the drainage system and the location of development within the catchment. The features required of methodologies for the generation of design information for flood alleviation in urbanizing catchments are presented and the degree to which existing design methods fulfil these requirements is examined. Flood-frequency techniques may be used to give estimates of post-urbanization flood levels, but hydrograph methods are necessary for the design of flood balancing works. A need still exists for a model capable of estimating the effects of non-uniform development and subcatchment flood storage, and also capable of yielding reliable estimates of flood frequency from storm frequency.

INTRODUCTION

The change in catchment response induced by urbanization is one of the most dramatic of man's impacts on the hydrological cycle. The basic effects of urbanization on catchment hydrology, both with respect to surface water drainage and the water supply distribution system, have long been recognised. A good account has been given by Savini and Kammerer (1961). This paper is concerned only with the changes in flood flow regime, how the rainfall-runoff process and the flood-frequency distribution are modified by urban development.

Urbanization changes the basic rainfall-runoff mechanism. The rendering impervious of previously permeable

land surfaces inhibits infiltration and groundwater
recharge, whilst the corresponding reduction in surface
roughness reduces both surface retention and depression
storage. As a result there is an increase in the percentage of rainfall that forms storm runoff. Furthermore the
alterations and improvements to the existing drainage network replace the relatively slow natural response with a
more rapid urban response. The combination of an increase
in storm runoff and a more rapid response is to produce
more runoff in less time, thus increasing flood potential.

The increase in flood potential may not be constant for
each rainfall event, but may vary with storm characteristics and antecedent conditions. On natural catchments,
more severe floods tend to occur when the incident rainfall intensity greatly exceeds the infiltration rate,
either because the rainfall is very intense or because the
catchment is already wet and of correspondingly reduced
infiltration capacity. Under such conditions, the response
of pervious surface may be similar to that of impervious
surfaces, and thus the effect of an increase in impervious
area would be less marked. Furthermore, on urban catchments, more severe floods may exceed the capacity of the
artificial drainage system causing localised ponding and
thus a reduction in flood levels downstream. Taken
together, these two effects may cause a reduction in the
flood frequency growth rate when compared with pre-urban
conditions, with smaller, possibly even negative, increases
being evident for the more severe, less frequent floods.
On the other hand, urban encroachment onto the natural
flood plain reduces the available overbank storage, thus
reducing the catchment's ability to attenuate over-bankfull
discharges. This may to some extent offset any reduction
in the flood frequency growth rate.

Therefore, the basic effect of urbanization on the
rainfall-runoff process is to increase storm runoff and to
increase the rapidity of response, while the magnitude of
each increase will depend on catchment characteristics,
storm characteristics, and antecedent conditions. It is
often assumed that the increases in storm runoff and in
rapidity of response arise from separate causes, the
increased storm runoff solely from the increased impervious
area and the increased rapidity of response solely from the
improved drainage system. This may be a good first
approximation, but each effect is a complicated interaction
of the increased impervious area, the improved drainage
system and the location of the developed area within the
catchment.

On natural catchments surface runoff occurs when the
incident rainfall intensity is greater than the infiltration rate and when the interflow discharge exceeds the
capacity of the upper soil layers which become 'wet' and
of zero infiltration capacity. Interflow is therefore
forced to emerge as surface flow. During rainfall, the
interflow discharge increases causing the 'wet' areas
adjacent to the drainage system to expand into the more
remote dryer areas. After the rainfall has ceased, the
'wet' areas will drain and contract. Catchment runoff can
then be considered as the joint effect of the emergent

interflow, approximately 100% runoff from the 'wet' areas,
the rainfall excess from the damp areas around the 'wet'
areas and approximately zero runoff from the 'dry' areas
beyond.

In this context there are five processes which influence
how building activity, the location of impervious surfaces,
and whether or not these are directly connected to the
drainage system affect the percentage of rainfall that goes
to runoff. First, building activity itself may increase
percentage runoff. Construction processes disturb the
natural top soil. With recompaction, reinstatement, and
turfing the infiltration capacity of the upper soil layers
may be much reduced. Furthermore, interflow paths may be
blocked or destroyed, causing any interflow to 'well up'
onto adjacent impervious areas, and thence into the artificial drainage system. Second, the introduction of
impervious surfaces may increase percentage runoff. However, if the new impervious areas are located in areas that
were previously 'wet' the change in percentage runoff may
be less than if the impervious areas are located in
previously 'dry' areas. Third, the introduction of an
improved drainage system may increase percentage runoff.
Whereas the natural drainage system may vary in length and
density with season and soil conditions, the artificial
drainage system is more constant and may reach areas, both
pervious and impervious, that previously did not contribute
to surface runoff. This effect is more evident during the
dryer seasons and contributes to a change in the seasonal
distribution of flood flows. Moreover, the introduction of
impervious surfaces without improvements to the drainage
system may not significantly affect percentage runoff, the
impervious area runoff soaking away on adjacent pervious
surfaces. Fourth, beside changing the overall percentage
runoff, urbanization changes the time distribution of
average loss rates. Runoff from the new impervious surfaces
may begin almost immediately, whereas, on the original
pervious surfaces, some rainfall was lost as infiltration
before runoff could begin. As a result average loss rates
may be more evenly distributed during the storm than before.
It may however be better to consider paved and unpaved
losses separately. Last, whereas the introduction of
impervious surfaces and the improvements to the drainage
system may cause a general lowering of groundwater levels
and hence a reduction in catchment wetness, the introduction
of a water supply and effluent disposal system may have the
opposite effect, with distribution losses and garden and
municipal irrigation increasing catchment wetness and
percentage runoff.

The increase in rapidity of response depends on the
interaction of the increased impervious area, the improvement and extensions to the drainage system, and the location
of development within the catchment. The more rapid
response arises from increased velocities of both surface
and channel flow. In small catchments, surface flow
represents a significant proportion of total flow time, and
thus, the introduction of impervious surfaces, by increasing
the velocity of surface flow, may alone represent a significant proportion of the increase in rapidity of response.

Conversely, in large catchments, channel flow represents a larger proportion of total flow time and the major increase in rapidity of response may be due to channel improvement. Thus the relative significance of impervious area and channel improvement on rapidity of response will vary with catchment size. Since urban response is more rapid than the corresponding rural response, non-uniform urban development within a catchment will alter the relative flow times of the response from different parts of the catchment. If the urban development is at a point remote from the catchment outfall, the quicker urban response may arrive at the outfall at the same time as the slower response from the natural areas nearer the outfall, thus yielding a reinforced peak discharge. Conversely, if urban development has taken place near the outfall, the quicker urban response may have passed the outfall before the natural response from the more remote areas has arrived, thus yielding a lower or even double peaked hydrograph. However, such downstream urban development may not cause concern at that design point, it may cause peak reinforcement at points further downstream. The lower response time of urban basins also means that they are able to respond more fully to shorter duration, higher intensity rainfall bursts than when in their rural condition. The urban peak flood of any specified probability will therefore tend to be caused by a shorter, more intense, peakier rainstorm. Thus, if a rainfall-runoff model plus design storm approach is used to estimate the change in flood frequency with urbanization, a different design storm should be used for pre- and post-urban conditions. Also, since in this country peakier rainstorms tend to become more frequent with the increase in convective rainfall during the summer, and since with increased urbanization antecedent wetness conditions have a reduced effect on catchment response, there is a progressive tendency for the flood season to move from the winter to the summer.

The effects of urbanization discussed above concern only changes in the flood runoff processes and it is usually assumed that the fainfall process is not affected by urban development. There is however growing evidence of local climate change with urbanization (Changnon, 1976), and Atkinson (Chapter 11) has reviewed evidence for increases in precipitation in urban areas. Such increases are usually neglected in estimating flood magnitudes since, as large cities take many years to develop, any early effects will already be incorporated in the local depth-duration-frequency data.

HYDROLOGICAL MODELLING IN URBANIZING CATCHMENTS

The development of a model to estimate the effect of urbanization on flood flow consists of three phases: the identification of the effect of urbanization; the choice of a suitable model and model parameters to describe the effect; and the generalisation of the model to allow its application to an ungauged catchment.

The identification of the effect of urbanization among other effects due to, for example, variation in rainfall or antecedent conditions, requires a high standard of data. To model the effect of urbanization on flood frequency long records from catchments in stable condition of urbanization are required, whereas to model the effect on the rainfall-runoff process good synchronised rainfall-runoff records at short time intervals are needed. Two sorts of study have emerged. In the first, a catchment is monitored throughout an urbanizing period with possibly a nearby rural catchment being monitored as a control. Thus information is obtained directly on the change in rainfall-runoff response, but usually, because of the shortness and non-stationarity of the record, no information on the change in flood frequency is available. Moreover, general trends can only be identified when the results from many such studies are collated. In the second type of study, the rainfall-runoff and flood-frequency characteristics of several catchments, both urban and rural, from within a region are compared, and any differences related to catchment characteristics including the degree of urbanization. Results, however, need careful interpretation to ensure that urbanization effects have been adequately separated from those of other catchment characteristics.

The choice of a model and model parameters to describe the effect of urbanization depends on the way the model is to be used. The need for a model to estimate the effect of urbanization on flood flows exists at two levels of sophistication. First, for planning, a simple model relating flood levels to urban development is needed to allow the evaluation of alternative development schemes. Such a model should be 'desk-top' or interactive computer based, require only very basic data input such as is available at the planning stage, and preferably be able to account broadly for the effects of inhomogeneous development and flood control structures. Second, for major drainage design a more detailed model is required to simulate accurately the response of the urbanizing catchment to enable the detailed design requirements of flood alleviation works to be specified. Such a model will have to cope with the downstream effects of inhomogeneous development and flood control structures, and the data requirements and computational complexity required will almost certainly mean the model will have to be computer based. Furthermore the model should be able to span the gulf between urban and rural hydrology and be capable of producing design flood information for any level of development.

Traditionally, rural flood hydrology has evolved along two lines: the statistical flood-frequency approach, based on long records of observed flood flows; and the deterministic rainfall-runoff modelling approach based on a concept of how the catchment converts rainfall to runoff. The two approaches are complementary in that the first provides an estimate of flood level at a specified probability while the second provides an estimate of both flood level and hydrograph shape due to a design storm of specified probability, though seldom have the probabilities

of derived hydrograph and design storm been related. With
these two approaches, it has been possible to decide
whether the better alternative for flood control is to
contain the flows by channel improvement and levée construction or reduce the flows by reservoir storage and
attenuation.

By contrast, urban flood hydrology has evolved almost
exclusively along the deterministic rainfall-runoff
modelling approach; the urban hydrological system lends
itself well to the deterministic approach since the flow
is predominantly over plane surfaces and in gutters, pipes
and channels, and as such can be analysed by the classical
laws of hydraulics. The virtual absence of the flood
frequency approach can be attributed to the lack of
sufficient long records from urban catchments in fixed
stages of development. Urban flood frequency is thus
usually estimated from a deterministic model, assuming the
resultant design flood to be of the same frequency as the
design storm. This assumption, however, has seldom been
tested for the particular model used. A reliable technique for estimating urban flood frequency is required if
the design level of flood protection is to be determined
by economic analysis. Furthermore, in an urban environment, surface water has tended to be considered a nuisance
to be conveyed swiftly from the catchment by a sufficient
drainage system. Thus traditional urban drainage design
methods give an estimate only of peak flow, which the
drainage system is then designed to pass. More recently,
with rapid urbanization and new town development, it has
been realised that passing on the increased flood waters
simply compounds the problem downstream, and thus may not
represent the optimal solution on economic or environmental
grounds. Planning authorities now tend to specify that
development should not alter the existing flood-frequency
distribution beyond prescribed limits. The concept of
Blue-Green development has evolved, whereby small flood
control reservoirs and temporary storage areas, designed
to both 'balance' the increased flood potential due to
urbanization and to replace some of the overbank storage
lost to flood plain development, are sited in 'linear
parks' of designated public open space or parkland along
the natural watercourses. Thus not only is urban flooding
alleviated but also the urban environment is enhanced. For
the design of such flood alleviation works a complete
design hydrograph is a vital prerequisite.

The choice of suitable model parameters to describe an
urban development is another problem. Percentage urban
area or percentage paved area are readily defined from maps
or aerial photographs, but the choice of parameters to
define channel improvement and the distribution of urban
development is less obvious. Furthermore, since drainage
improvements and increases in paved area tend to occur
simultaneously, parameters describing these changes tend to
exhibit high statistical intercorrelation. This may lead
to difficulties if the model is to be used to estimate
changes due to either cause separately.

The final stage of model development is the generalisation of the results. To be able to apply the model to

estimating the effects of urbanization in an ungauged
catchment or future conditions in a gauged catchment, it
must be possible to estimate model parameters from catchment characteristics. The more empirical methods are
usually fitted to observed rainfall-runoff data and the
chosen optimum parameters regressed on catchment characteristics. The use of such equations to estimate model
parameters for conditions not truly comparable to those on
which the equations were derived may lead to errors. This
caution applies particularly to urbanizing catchments since
urbanization may change the natural areal distribution of
contributing areas and overland flow lengths, which before
conformed to some regional statistical norm. To cite two
examples of a more unexpected nature, Crippen (1965)
investigated changes in unit hydrograph shape due to
urbanization in Sharon Creek, California, where a golf
course was situated just upstream of the gauging station.
He found that the post-urban unit hydrograph had a shorter
recession limb and a higher peak discharge due to the
faster response from the more remote urban areas, but that
time to peak had not changed since the urban response
simply reinforced the existing peak from the golf course
area. Wallace (1971) investigated the effect of urbanization on runoff from the Peachtree watershed in Atlanta,
Georgia. Early urbanization had occurred in hilltop areas,
while later urbanization occurred in valley areas. The
period of record examined only covered the later urbanization. Comparison with a control catchment suggested
that a constant increase in percentage runoff had occurred
due to hilltop urbanization in previously dry areas.
Valley urbanization during the period of record showed, for
winter events, no significant effect on either volume of
runoff or unit hydrograph shape, but for summer events, a
three-fold increase in volume of runoff. Such results
suggest the need for a distributed model to estimate the
effects of urbanization, or at least, more careful definition of the range of applicability of more simple models.

With the more physically based models (eg. the Stanford
Watershed model, the EPA Storm Water Management Model and
the MIT catchment model) in which the physics of catchment
response is more realistically modelled, model parameters
tend to correspond to specific catchment characteristics
(eg. depth to impermeable soil layers, overland and channel
flow lengths and roughnesses). It is then often assumed
that the effects of urbanization may be adequately
estimated by the intuitive pro-rata adjustment of model
parameters. However, since the model is not exactly
equivalent to the real world, this may lead to errors, and
it would be wise therefore to test this assumption by
fitting the model to observed pre- and post-urbanization
records. Unfortunately in few instances has this been done.

EXISTING DESIGN METHODS

To what extent do existing methods fulfill the needs of a
model to estimate the effects of urbanization? Carter
(1961) proposed a method for determining the effect of

urbanization on the mean annual flood. Anderson (1970)
extended Carter's work and included the estimation of the
urbanized growth curve. In the method, mean annual flood
is determined from the following equation (Anderson, 1970)
based on 44 catchments in the Washington-Northern Virginia
area ranging in size from 1 to 1476 km^2 and in percentage
impervious from 0 to 30.

$$\bar{Q} = 230 \text{ K.A}^{.82} \text{ T}^{-.48}$$

where $\bar{Q}$ is mean annual flood (ft^3/s), K is a co-efficient
of imperviousness defined as 1.00 + 0.015 I, I is
percentage of catchment impervious, A is catchment area
(mi^2), and T is lag time (hr) defined as the time from
centroid of rainfall excess to the centroid of direct
runoff.

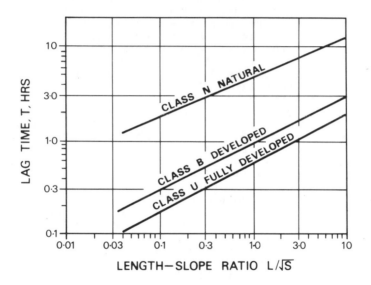

Figure 1. Lag time v. basin ratio (after Anderson, 1970).

For the ungauged situation, lag time is defined
graphically (Figure 1) in terms of a channel length-slope
ratio, L/$\sqrt{S}$ (where L is distance (mi) along the main channel
from outlet to divide and S is slope (ft/mi) of main channel
from .1L to .85L upstream of outlet), and the degree of
drainage improvement. Regression equations are presented
for natural conditions, partially developed conditions -
sewered tributaries but natural main channels, and fully
developed conditions - fully sewered. The mean annual
flood is then scaled up to a T-year flood using the urbanized growth curve, which Anderson (1970) assumed could be
found by interpolation using the percentage imperviousness
between the natural growth curve, found by regional analysis
of rural catchments, and the fully urbanized growth curve,

which was assumed to equal the rainfall growth curve.
Sauer (1974) combined Anderson's (1970) approach for the
urbanized growth curve with Leopold's (1968) graph (Figure
2) of the factor, R, by which percentage imperviousness
and percentage channelisation increases the rural mean
annual flood, $\bar{Q}_R$. He presents a simple explicit formula
for the urbanized T-year flood, Q_{TU}, in terms of the rural
R-year flood, Q_{TR}, and the corresponding T-year factor
from the rainfall growth curve, $P_T/\bar{P}$.

$$Q_{TU} = \frac{P_T}{\bar{P}} \cdot \frac{7 \bar{Q}_R (R-1)}{6} + \frac{Q_{TR} (7-R)}{6}$$

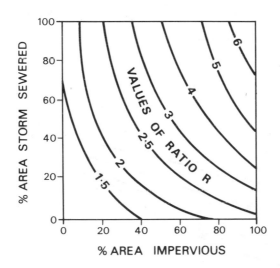

Figure 2.   Urban adjustment ratio, R, for mean annual
    flood (after Leopold, 1968).

Several workers have derived regional regression equations
for mean annual and various return period floods in terms
of catchment characteristics.
    In the U.K., NERC (1975b) present the results from a
countrywide Flood Study. Regional regression equations are
presented for mean annual flood in terms of catchment
characteristics, and also regional and national growth
curves are presented for the ratio of T-year to mean annual
flood based on the general extreme value distribution (GEV).
Records from 420 catchments were used in the study, but
only 19 catchments were more than 20% urbanized most of
which were in the London area. Furthermore only a rather
crude, but easily obtainable, index of urbanization was
used, namely the proportion of 'grey area' on a 1:63360
O.S. map. As a result, urbanization only appears as a
significant catchment characteristic in the Thames, Lee and
Essex region. Further work on the Floods Study data

(Packman et al., 1976) has resulted in a compromise national factor by which to multiply the rural mean annual flood in order to take account of urbanization:

$$\bar{Q}_u/\bar{Q}_R = (1 + URBAN)^{1.8}$$

where URBAN is the fraction of the catchment under urban development. Furthermore fitted values of GEV parameters have been related to urbanization, and thus a tentative national urbanized growth curve (Figure 3) has been drawn up to allow the estimation of the T-year urbanized flood in terms of the natural mean annual flood and degree of urbanization. Such flood-frequency techniques however give only a peak discharge value and are unable to account for inhomogeneous development. They may be used at a planning stage to give an indication of the expected changes, but will be unable to give an estimate of the effect of any flood alleviation measures.

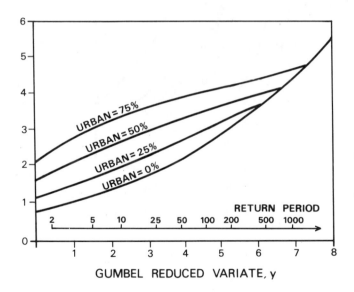

Figure 3. National growth curve for various degrees of urbanization (after Packman et al., 1976).

In order to be able to predict the post-urbanization hydrograph shape, unit hydrograph theory has been applied quite extensively. If the effective rainfall-direct runoff process can be considered linear then the unit hydrograph, by being defined for a unit depth of rainfall in a unit duration, can be considered a function of the catchment routing process only. Thus, assuming no change in the rainfall process, the change from pre- to post-urban rainfall-runoff response can be estimated by considering the changes in unit hydrograph and rainfall losses separately. Far less work has however been done on the effect of urbanization on rainfall losses than on unit hydrograph shape. Snyder's (1938) synthetic unit hydrograph technique has been used extensively to estimate changes in unit hydrograph peak and time to peak due to urbanization, but the problems of defining coefficient values has now lead to its virtual obsolescence. Van Sickle (1963-64), mindfull of Carter's (1961) work, presented graphs for unit hydrograph peak and time to peak in terms of slightly different length-slope factor and the degree of development of the drainage system. However, the first method to gain general acceptance was that of Espey et al. (1965, 1969). Based on data from 33 urban and 17 rural watersheds in Texas, they derived the following equations for 30 minute unit hydrograph peak and rise time:

$$\text{URBAN CATCHMENTS} \quad Q = 3.5 \times 10^4 \, A^{1.0} \, T_r^{-1.10}$$

$$T_r = 16.4 \, \Phi \, L^{.32} \, S^{-.049} \, I^{-.49}$$

$$\text{RURAL CATCHMENTS} \quad Q = 8.25 \times 10^4 \, A^{.99} \, T_r^{-1.25}$$

$$T_r = 2.68 \, L^{.22} \, S^{-.30}$$

where Q is unit hydrograph peak (ft^3/s), A is area (mi^2), T_r is rise time (minutes), L is main channel length (ft), S is main channel slope (ft/ft), I is percentage impervious cover and Φ is a channel roughness factor determined from Table 1.

These equations have been used by many workers and have proved capable of estimating pre- and post-urbanization unit hydrograph parameters in the USA and the UK. The choice of a suitable value for Φ is, however, a large area of subjectivity.

Rao et al. (1972) presented a conceptual model instantaneous unit hydrograph (IUH) method based on about 200 storm events from 5 rural and 8 urban catchments in Indiana and Texas. They concluded that for catchments of less than 5 mi^2 the linear reservoir model was adequate. The following equations were used to define the IUH, h(t):

$$h(t) = \frac{1}{K} e^{-t/K}$$

$$K = .887 \, A^{.49} \, (1+U)^{-1.683} \, P_e^{-.24} \, T_R^{.294}$$

where A is area (mi^2), U is impervious fraction, P_e is the volume of rainfall excess (in), and T_R is the duration of rainfall excess (hr). For catchments in excess of 5 mi^2 they preferred the Nash cascade model in which the IUH is given by:

$$h(t) = \frac{e^{-(t/K_N)} \, t^{(n-1)}}{K_N^n \, \Gamma_n}$$

$$K_N = .575 \, A^{.389} \, (1+U)^{-.622} \, P_e^{-.106} \, T_R^{.222}$$

$$n = 1.445\ A^{.069}\ (1+U)^{-1.040}\ P_e^{-.161}\ T_R^{.149}$$

Note that each set of equations assumes the unit hydrograph shape is constant during the event, but that it will vary between events with different values of P_e and T_R - a quasi-linear analysis. The equations do not however predict such large effects as have been observed elsewhere, and their validity should be checked before they are applied in other areas.

Table 1: Φ Classification for Espey et al. (1965, 1969) equations. $\Phi = \Phi_1 + \Phi_2$

Degree of channel improvement	Φ_1
Extensive channel improvement. Storm-sewer system and closed conduit channel system	0.6
Some channel improvement. Storm sewers, but mainly cleaning and enlargement of existing channel.	0.8
Natural channel conditions	1.0
Degree of channel vegetation	Φ_2
No channel vegetation	0.0
Light channel vegetation	0.1
Moderate channel vegetation	0.2
Heavy channel vegetation	0.3

In the U.K., Hall (1973) presented a method based on the analysis of 63 storm events from 5 catchments in the Crawley, Sussex area. The method is based on a graphically defined one-hour dimensionless unit hydrograph shape (Figure 4), which may be dimensionalised by scaling both the discharge and time scales by the one parameter, lag time. Lag time is defined as the time from the centroid of the unit storm to the centroid of the unit hydrograph, and for the ungauged situation may be estimated from a graph similar to Anderson's (1970) (Figure 5).

The USDA Soil Conservation Service (1975) present methods for determining both losses and synthetic unit hydrographs for urbanizing catchments. Losses are determined using the SCS soil-cover-complex method, in which runoff volume, Q, is given in terms of rainfall volume, P, and potential abstraction, S, by:

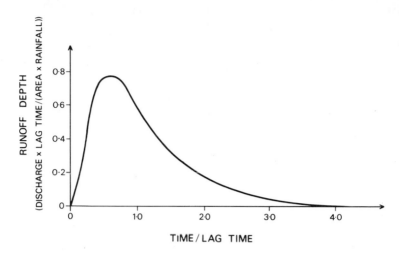

Figure 4. Non-dimensional one hour unit hydrograph
(after Hall, 1973).

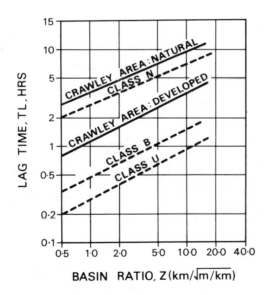

Figure 5. Lag time v. basin ratio (after Hall, 1973
and Anderson, 1970).

$$Q = (P-0.2S)^2/(P+0.8S)$$

$$S = (1000/CN) - 10$$

where, CN, the 'curve number' depends on soil group, land
use, and antecedent conditions. Soil group information for
locations in the USA are given, and Table 2 shows how curve

Table 2.   Soil Group Information for the USDA Soil
     Conservation Service synthetic unit hydrograph method.

LAND USE DESCRIPTION		SOIL GROUP			
		A	B	C	D
CULTIVATED LAND:	with water conservation treatment	62	71	78	81
	without water conservation treatment	72	81	88	91
PASTURE OR RANGE LAND:	Poor condition	68	79	86	89
	Good condition	39	61	79	80
MEADOW; Good condition		30	58	71	78
WOODLAND AND FOREST:	Thin stand, poor cover, no mulch	45	66	77	83
	Good cover, no brush or litter	25	55	70	77
OPEN SPACES, PARKS, ETC:	75% or more grass	39	61	74	80
	50% to 75% grass	49	69	79	84
COMMERCIAL AND BUSINESS AREA: (85% IMPERVIOUS)		89	92	94	95
INDUSTRIAL DISTRICTS: (72% IMPERVIOUS)		81	88	91	93
RESIDENTIAL:	8 houses per acre (65% impervious)	77	85	90	92
	4 houses per acre (38% impervious)	61	75	83	87
	2 houses per acre (25% impervious)	54	70	80	85
PAVED ROADS, CAR PARKS, ROOFS, DRIVEWAYS ETC.		98	98	98	98

SOIL GROUPS: A - Low runoff potential, deep well drained sands and
                    gravels
               B - Moderately good depth and drainage, moderately
                    coarse texture
               C - Moderately poor depth and drainage, moderately fine
                    texture
               D - High runoff potential, shallow poorly drained clays

numbers may be obtained from land use. For application in
the UK, however, soil group information is not readily
available. The SCS triangular synthetic unit hydrograph
due to 1 inch of rainfall in duration D hours is given by:

$$q_p = 484 \ A/t_p$$

$$t_b = 2.67 \ t_p$$

$$t_p = D/2 + T_L$$

where q_p is the peak of the unit hydrograph (ft^3/s), A is
the catchment area, t_p is the time to peak of the unit

hydrograph (hr), t_b is the time base of the unit hydrograph (hr), and T_L is the basin lag, defined as the time from the centroid of rainfall excess to time of peak direct runoff.

In small (less than 8 km^2) ungauged natural catchments, T_L can be estimated using the following equation:

$$T_L = \frac{L^{0.8} (S+1)^{0.7}}{1900 \; Y^{0.5}}$$

where L is the hydraulic length of the catchment (ft), S is the potential abstraction referred to above, and Y is the average land slope (%). The Soil Conservation Service present graphs (Figures 6 and 7) of adjustment factors for natural lag time to account for percentage impervious and channel improvement.

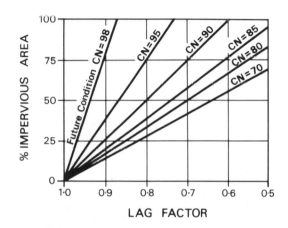

Figure 6. Impervious area lag factor (after USDA, 1975).

The NERC (1975b) UK Flood Studies Report presented techniques for estimating rainfall losses and synthetic triangular unit hydrographs. Rainfall losses were estimated assuming a constant percentage runoff given by:

$$PR = 0.22 \; (CWI-125) + 0.10 \; (P-10) + 95.5 \; SOIL + 0.12 \; URB$$

where CWI is an index of catchment antecedent wetness (see NERC, 1975b, vol. 1, sections 6.4.4 and 6.7.6), P is the rainfall depth, SOIL is an index of soil type (see NERC, 1975b, vol. 1, section 6.5.7) and URB is percentage urban. This equation is based on the analysis of 1447 flood events from 130 catchments, of which 24 were greater than 5% urbanized. Further work at the Institute of Hydrology (Packman et al., 1976) has indicated that while the above equation gives a best estimate of percentage runoff under any fixed conditions of urbanization, if the rural percentage runoff, PR_R, is known, the urbanized percentage runoff, PR_u,

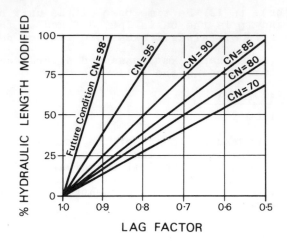

Figure 7.   Main channel improvement lag factor (after
    USDA, 1975).

may be better estimated by:

$$PR_u = PR_R - 6.87 \text{ SOIL} + 0.12 \text{ URB} + 2.0$$

The NERC (1975b) triangular unit hydrograph is defined by:

$$Q_p = 220/Tp$$

$$TB = 2.52 \ T_p$$

where Q_p is unit hydrograph peak discharge (m^3/s per
100 km^2), Tp is unit hydrograph time to peak (hr) and TB is
unit hydrograph time base (hr).   For the ungauged situation,
an estimate of Tp is given by:

$$Tp = 46.6 \ L^{0.14} \ S^{-.38} \ (1 + \text{URBAN})^{-1.99} \ \text{RSMD}^{-0.42}$$

where L is main channel length (km), S is main channel slope
between points .10L and .85L upstream of outlet (parts per
1000), URBAN is urbanized fraction and RSMD is an index of
climate – the net one day once-a-year effective rainfall.
Further work at the Institute of Hydrology (Packman et al.,
1976) has indicated that if the natural unit hydrograph
time to peak, Tpr, is known, the urbanized time to peak,
Tpu, may be better estimated from:

$$Tpu/Tpr \quad 1 \quad (1 + \text{URBAN})^{-1.95}$$

All the above unit hydrograph techniques can be used to
provide a complete design hydrograph.   The NERC (1975b)
technique is further able to estimate the frequency of that
design hydrograph rather than simply assume it to be the
same as the storm input.   None, however, are able to account
for the effects of inhomogeneous development.   To account

168

for the effects of inhomogeneous development, several
models have been developed in which catchments are simulated
as an assemblage of subcatchments of uniform characteristics. Unit hydrographs may then be used to define subcatchment response, and subcatchment response may be routed
to the main catchment outlet. Such approaches are generally
too complex for easy manual solution, and are usually prog-
rammed for computer solution.

Evelyn et al. (1970) developed such a model from
Narayana and Riley's (1968) model of the urbanizing Waller
Creek Watershed in Texas. This latter model is based
essentially on unit hydrograph theory, with rainfall loss
and unit hydrograph lag time parameters being related to
percentage imperviousness and a characteristic impervious
length factor which is the ratio of the area weighted mean
distance of impervious area from the basin outlet to the
length of the basin. As such the model already makes some
allowance for inhomogeneous development. Evelyn et al.
extended this allowance by using Narayana and Riley's model
to estimate the response from subcatchments within Waller
Creek and routing these down the main channel using a
linear reservoir. The regression equations for subcatchment parameters cannot however be used for other
catchments since they contain no catchment characteristics,
only urbanization characteristics.

The US Army Corps of Engineers (1971) and the USDA Soil
Conservation Service (undated) have developed similar
distributed unit hydrograph models (HEC-1 and TR-20
respectively), which determine catchment response, as the
joint effect of several subcatchment responses. Rainfall
input to either model may be observed data or standard
design storms. HEC-1 determines rainfall losses either as
an initial loss plus continuing loss or as a function of
accumulated loss and rainfall intensity. TR-20 determines
rainfall excess using the soil-cover-complex method. The
subcatchment unit hydrographs may be given as input or
estimated by Clark's (1945) method in HEC-1, while TR-20
uses a standard dimensionless form. Channel routing in
HEC-1 uses either the Muskingham or Straddle-Stagger
techniques while TR-20 uses the Convex Method. Reservoir
routing in each model uses the normal level pool equations.
Either model may be used to estimate the effects of urban
development and flood alleviation works, using previously
published results to adjust unit hydrographs and loss
rates in the separate subcatchments. The USDA Soil Conservation Service (1975) present tabular approximations to the
routed subcatchment hydrographs that would be obtained by
applying TR-20 to various subcatchment lag-times and channel
flow times. Using these tables together with the rules they
present for adjusting soil-cover-complex curve numbers and
lag times to account for urbanization, an approximate design
hydrograph at the basin outlet can be determined, simply by
superimposing the routed subcatchment hydrographs after
having scaled them up for subcatchment area and volume of
runoff.

Dempster (1974) used a modified version of the USGS
model (Dawdy et al., 1972), in which pervious and impervious
areas were handled separately, with separate unit hydro-

graphs determined using Clark's (1945) method. Effective
rainfall separation in the model uses an antecedent moisture accounting model and an infiltration model, the
parameters of which need fitting to the specific catchment.
No equations for estimating these parameters for the
ungauged situation, or for estimating the effect of urbanization of these parameters are presented. Such models
assume subcatchment routing may be adequately represented
by linear, unit hydrograph techniques, thereby keeping the
calculation procedure simple. However, since the models
have already become computer-based to aid the calculation
management, the computational power of the computer may
also be used to solve more realistic non-linear subcatchment routing techniques.

One of the simpler of the non-linear techniques is
Laurenson's (1964) non-linear reservoir model, which Aitken
(1975) has developed for urban and rural catchment applications. The model is a rainfall excess routing model only,
but Aitken advises that rainfall losses for pervious and
impervious areas should be estimated separately. The model
divides the catchment into 10 sub-areas delineated by lines
of equal travel time to the outlet. Sub-area routing uses
10 equal non-linear reservoirs, each given by:

$$q = BS^{.715}$$

$$B = 0.581 \; A^{.436} \; (1.0 + U)^{-2.740} \; S_c^{-.339}$$

where q = sub-area outflow (m^3/s), S = sub-area storage (m^3),
A = total catchment area (km^2), U = fraction of total
catchment urbanized and S_c = main channel slope (%).

Several more physically based models have also been
developed, using the hydraulic flow equations. Lanyon and
Jackson (1974) developed the Chicago Flow Simulation
Program, a model designed for use at the planning stage and
needing only simple catchment information. Rainfall losses
are determined from rainfall intensity, overland flow
length, soil moisture, and certain regional empirical
constants. Surface runoff and sewer routing use linear
approximations to the kinematic flow equations, and channel
routing uses the kinematic flow equations. The model has
been applied in Chicago and Illinois, but values for the
regional constants should be found before the model is
applied in other areas.

The Stanford Watershed Model has been applied extensively to urbanizing catchments, particularly in its
Kentucky version (James, 1972). Land phase processes
simulated by the model consist of interception, infiltration and depression losses, and overland flow routing
based on an empirical relationship between depth and discharge. Channel phase processes are simulated by Clark's
(1945) method. Individual calibration to the particular
catchment is necessary since reliable methods of estimating
model parameters from catchment characteristics have yet to
be developed. This limits its applicability for ungauged
catchments. The model operates in continuous time using
observed long records of rainfall and evaporation to produce

a synthetic flow record suitable for flood-frequency analysis. Thus, although the model requires more data for both calibration and operation purposes, the problem of specifying design antecedent catchment characteristics is avoided. Once fitted, the effect of urbanization is claimed to be reasonably estimated by intuitive variation of model parameters (James, 1965; Ligon and Stafford, 1974; Durbin, 1974).

Two other physically based models have been developed for use in urban and urbanizing catchments, the MIT model (Harley et al., 1970) and the EPA Storm Water Management Model (Metcalf and Eddy et al., 1971). Each conceptualises the catchment into an assemblage of rectangular subcatchments, calculating overland and channel flow hydrographs using kinematic formulations of the hydraulic equations of non-steady flow. The models may be fitted using regional constants for rainfall loss parameters and runoff routing parameters obtained directly from physical catchment characteristics. However, catchment data requirement is quite extensive which may mean the models are inapplicable at the planning stage. Furthermore the computer requirement of each is large and thus users have tended to fit simpler models or simpler catchment configurations to the results of the full model to allow more economic simulation of a larger number of events. However the ability of the models to estimate the effect of urbanization on flood flow regime has not been tested against real data.

POSSIBLE FUTURE LINES OF RESEARCH AT THE INSTITUTE OF HYDROLOGY

The above discussion indicates that techniques, albeit of a tentative nature, have been developed to allow for the effect of urbanization on flood peaks and hydrograph characteristics under UK conditions. No technique, however, has been developed to allow for non-uniform urban development. The research programme at the Institute of Hydrology is therefore concerned with improving the reliability of the existing techniques and developing a model to account for non-uniform development.

Initial work under this research programme (Packman et al., 1976) has abstracted as much information on the effects of urbanization as was readily available from the existing data set gathered for the NERC Flood Studies Report (1975b). Since that data collection programme was completed (1969), 6 or 7 years of additional data have been collected, several stations have passed the 5 year minimum record length criterion set for the Flood Studies Report, and several new stations have been installed. This increased data set should allow more specific study of the effect of urbanization in order to improve the simple urbanization factors already obtained.

The Flood Studies Report recommended a unit hydrograph technique for catchment routing and both the Muskingham and Variable Parameter Diffusion techniques for channel routing. These may be combined in a manner similar to that of the US Army Corps of Engineers model, HEC-1, or the USDA

171

Soil Conservation Service model, TR-20, to produce a computer-based distributed unit hydrograph model of catchment response. Although a non-linear subcatchment routing technique might produce more realistic simulations of catchment response, for design purposes, linear techniques, based on the considerable volume of experience that exists, may be considered adequate, particularly if suitable design conditions can be specified. In effect, the effort spent in developing a non-linear technique may be better spent developing a linear model able to estimate floods of specified frequency. This does not deny the need for more realistic subcatchment routing, but indicates that, at present, improvement of existing linear methods represents a more fruitful area of research.

Once having developed a distributed unit hydrograph model of catchment response, the model could be run with a range of parameter values in order to develop a simple manual approximation method, similar to that given by the USDA Soil Conservation Service (1975) for their TR-20 model. Such an approximation method might be used at a planning stage to investigate alternative development strategies. As the approximation method and the full model would be largely comparable as regards specification of catchment information, the 'best' strategy chosen by the approximate method could be later analysed by the full model with a minimum of further effort.

With regard to the development of non-linear techniques, the prime necessity is not for new models but for the refinement and development of fitting methods for existing models. Considerable experience is being gained in the US with two models in particular, the Stanford Watershed Model and the EPA Storm Water Management Model. The suitability of these models for UK conditions should be investigated and ways of estimating model parameters from catchment characteristics developed.

ACKNOWLEDGEMENTS

This paper is presented by kind permission of Dr. J.S.G. McCulloch, Director of the Institute of Hydrology. Work at The Institute of Hydrology is supported by the Department of the Environment under Contract number DGR/480/38.

14

STORM FLOWS IN SUBURBIA: THE SOUTHAMPTON SMALL CATCHMENT EXPERIMENT

P.R. Helliwell and C.H.R. Kidd

Binnie and Partners, Artillery House, Artillery Row, London, SW1; and Institute of Hydrology, Wallingford, Oxfordshire.

ABSTRACT

Urban storm runoff involves an above-ground hydrological phase and a below-ground hydraulic phase. The investigation of the above-ground phase in two tiny suburban catchments in Southampton is described. A non-linear reservoir routing method is shown to be superior to the traditional time of entry method for estimating sewer inlet hydrographs for very small basins. A simulation experiment showed that this superiority may extend to large catchments too.

INTRODUCTION

Urban storm runoff is a two-stage process. Excess storm precipitation travels overground to collecting points beyond which flow is in pipes. There is no clear-cut interface between the two phases. For practical reasons the best point of division is the manhole in the sewer system. The overground phase therefore encompasses the conversion of storm precipitation into overland flow, its concentration into formal and informal channels, and its translation through short lengths of pipe from gully traps. The below-ground stage is concerned with combining and routing inlet hydrographs from the above-ground stage through the sewer system to the outfall. The above-ground stage is dominated by hydrological controls and processes. The below-ground stage is dominated by hydraulic considerations. Progress towards a better understanding of urban storm runoff phenomena can only be made if both these stages are studied. Urban runoff data previously collected in the U.K. generally relates to relatively large catchments in which both stages are important and the hydraulics of pipe flow may be the dominant problem. It is to these latter problems that the majority of researchers have directed their attention.

173

The work at Lordshill, Southampton, has been set up specifically to study the above-ground hydrological stage of the whole process. This is achieved by selecting sewered areas in which the modification to the surface runoff hydrograph in the sewer pipe is small. The areas selected are therefore very small, being less than one hectare in both cases. One of the components of any mathematical model of urban drainage deals with surface routing or the conversion of a net rainfall hyetograph into an inlet hydrograph. The traditional approach to this aspect, used in both the Rational Formula and the Road Research Laboratory Method (Watkins, 1962), employs a time of entry model which routes the net rainfall through a linear time-area diagram of base length equal to a specified period, normally 2 minutes.

The interface between the above-ground, hydrological phase and the below-ground, hydraulic phase, is not sharply delineated. Flow into down-pipes from roofs and into gully-traps along road-sides are the obvious points of division. It is only within the last few years that instruments capable of measuring flow into gully traps have been developed (Blyth and Kidd, 1977), and there is still a need for an instrument to measure drainage from individual roofs. On the other hand, equipment for measuring flow in channels is well-established, the main peculiarity in urban flow measurement being the need for accurate relative timing of the rainfall input and resulting stormflow. Time constants of the order of 2 minutes are involved, so that relative timing to better than half a minute is needed.

Urban drainage investigators in Britain have generally accepted that a simple time delay of two or three minutes is an adequate model of catchment response. Further advances will only be made by applying a more realistic model to the runoff process. Some success has been achieved with a single non-linear reservoir model (Kidd and Helliwell, 1976; Kidd, 1976).

A major difficulty is the prediction of total quantity of runoff. This is currently done in two stages. A depression storage sub-model is used which, unless satisfied by previous rainfall, must be filled before surface flow occurs. This is followed by a volume runoff sub-model, which assumes 100% runoff from paved and roofed areas and 0% from all other surfaces. The errors inherent in these assumptions, the uncertainty of catchment rainfall estimation for short time increments, the difficulty of measuring accurately flows in the long tail of the hydrograph and uncertainties in location of catchment boundaries present a major obstacle to reliable prediction of runoff volume. The prediction of the timing and shape of the response of the catchment is more accurate and reliable than the prediction of the scale of the response. Recent work at the Institute of Hydrology is aimed at this area (Stoneham and Kidd, 1977).

DATA COLLECTION

The small catchments in Lordshill, a new suburb of Southampton have been monitored for rainfall and runoff since June 1974 (Figure 1). During this time some 40 usable storm events

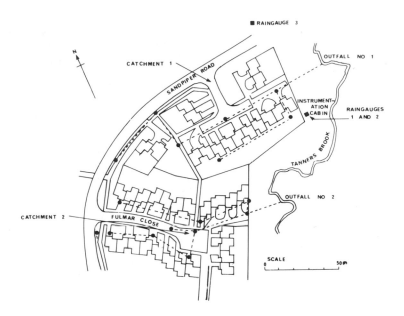

Figure 1. Plan of Lordshill catchments

have been recorded. Housing density is about 20 units per
hectare, the soils arc heavy clay, and the roofs are connected
to the storm sewer system.

Catchment No 1 drains a small close and a portion of an
estate road. The gross area is 8000 m^2, of which 3340 m^2 is
sealed. 2260 m^2 is paved, and 1080 m^2 roofed. Average ground
slope is 3%. There is difficulty defining catchment boundary
on the estate road. The topographic divide for unsealed
surfaces is remote. Catchment No 2 drains a close. It has
a gross area of 6000 m^2, of which 2520 m^2 is sealed. 1250 m^2
is paved and 1270 m^2 roofed. Average ground slope is 6%.

The runoff from the two catchments is monitored at their
respective outfalls using 229 mm U-shaped Venturi flumes
supplied by Arkon Instruments. The water-levels in the
stilling-basins are monitored using the Arkon air-purge system.
A duplex recorder, housed in the instrument cabin, records
traces from both flumes onto one chart. Chart speed is
76 mm/h. A theoretical calibration, which was checked in
the field using artifically-induced flows, was employed for
the conversion of water-level to discharge.

The rainfall is monitored using three gauges, the locations
of which are shown in Figure 1. Gauge Number 1 is a Casella
Tipping Bucket Raingauge (W5698) and is located on the roof
of the instrument cabin. Each tilt of the bucket is registered
inside the cabin on a Casella Receiver (W5705), with a chart
speed of 300 mm/day. Raingauge tips are simultaneously
recorded on the Arkon recorder by means of a third pen fitted
to the Arkon recorder. The benefits of this extension are
two-fold; firstly, it enables the rainfall and runoff events

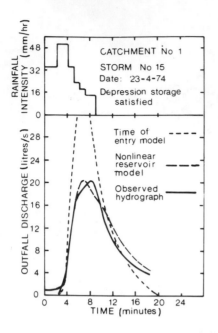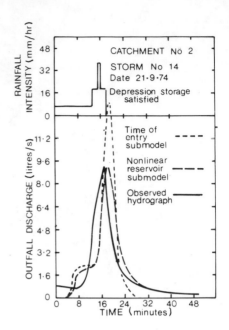

Figure 2. Simulation of historic rainfall-runoff events
    on Lordshill catchments

to be synchronised; and secondly, it provides more sensitive
definition of higher rainfall intensities by virtue of the
greater speed of the Arkon chart drive. Gauge Number 2 is a
Casella Snowdon Standard Raingauge, and is sited next to the
tilting bucket gauge on the cabin roof as a check on the
latter's performance. Gauge Number 3 is also a Casella
Snowdon Standard Raingauge. It is sited in a position which
conforms with British Meteorological Office recommendations.
Owing to its correct siting, it is employed as the reference
gauge, and the tipping bucket gauge is used to distribute
temporally the total rainfall depths measured by Gauge No 3.
It has been found that, on average, the roof gauges catch
less than Gauge Number 3.

THE IMPORTANCE OF THE ABOVE-GROUND PHASE

Analyses (Kidd and Helliwell, 1976; Kidd, 1976) using the
Lordshill data, have shown that a simulation which uses the
net rainfall as input to a non-linear reservoir provides a
good simulation of the surface routing process, whereas the
same model using a 2-minute time of entry does not. (Figure
2).

The equation of the non-linear reservoir is

$$S = Kq^n$$

$$\frac{ds}{dt} = i - q$$

where S is storage in the system, i is inflow to the system, q is outflow from the system, and n and K are parameters.

Suitable parameter values for the two Lordshill catchments have been found (Helliwell and Kidd, 1977) to be

n = 0.67

K = 5.2 for roofs

26.0 for paved areas.

The time of entry model has never been considered an accurate simulation tool, but was considered adequate for larger catchments where the predominant part of the routing process was believed to occur underground. It seems reasonable to suppose that the difference in model performance between a time of entry model and a more complex counterpart might decrease with increasing catchment area. It seems appropriate to attempt to gain some insight into the extent to which this latter hypothesis is true.

To this end, a number of rainfall-runoff simulations were performed using the two model types mentioned above, using a 2-hour 1-year return period rainfall event (Road Research Laboratory, 1963) over a series of hypothetical catchments of increasing size. The model used was the University of Southampton Urban Runoff Model, which is described in detail elsewhere (Kidd, 1976). It uses a surface routing submodel which can adopt either a non-linear reservoir or a time of entry mechanism as required. Pipe-flow routing uses the kinematic wave equations. The smallest of the series of catchments used is Lordshill Number 1, and the model parameters are those obtained in the earlier study. Figure 3 shows the rainfall input and the modelled outfall hydrographs for the 2-minute time of entry and the non-linear reservoir submodels respectively. As for the monitored events in the earlier study, the former produces a peak discharge which is greater and slightly earlier than the more realistic model.

For extending the comparison to larger catchments, Lordshill Number 1 was considered as an 89 m square unit draining at one corner. The next size (4-unit) catchment comprised four such catchments linked by a pipe network in the manner shown in Figure 4. All pipe slopes were 1½%, with diameters sufficient to accept peak discharges without surcharging. The next size (16-unit) catchment was synthesised in the same fashion, now using the 4-unit catchment instead of the 1-unit catchment. The same procedure was adopted for a 64-unit and 256-unit catchment - the last is a 205 ha area with a maximum pipe size of 1.22 m (48 inch). The final catchment is, then, a large homogeneous catchment having land-use characteristics which are the same as those for Lordshill Number 1. Such a catchment network is clearly not entirely realistic, but it does contain typical total lengths of appropriate sewer sizes.

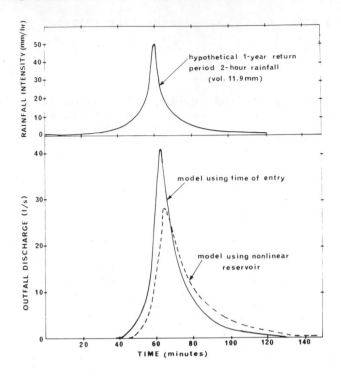

Figure 3. Model performance on Lordshill No 1.

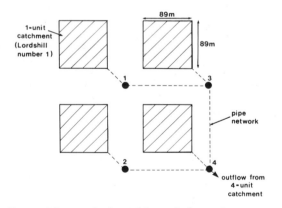

Figure 4. Formation of 4-unit catchment

Simulations were performed using the hydrograph resulting from one catchment as input to the catchment of next size up (e.g. the runoff hydrograph from the 4-unit catchment is the input hydrograph to the 16-unit catchment). The ratio of the time of entry modelled hydrograph peak value to the non-linear reservoir peak value was calculated and plotted against catchment area in Figure 5. As seemed likely, the error due to use of a simplified (time of entry) surface routing model

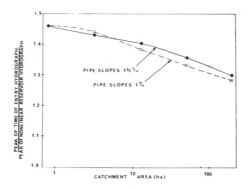

Figure 5. Effect of increasing catchment area

instead of a more realistic one does decrease with time but
not by as much as might have been expected. It was considered
that the decrease in the effect might be more marked for a
flatter catchment where the water would spend a greater time
below ground. The simulations were repeated using pipe slopes
of 1% in place of 1½%, but the decrease in the effect was
still not large (Figure 5).
     Generalised conclusions arising out of these analyses
must necessarily be tentative, because they concern the
application of a particular storm to a series of particular
hypothetical catchments using a rainfall-runoff model whose
performance has not yet been verified on large catchments.
However, indications are that above-ground routing is as
important as below-ground routing for a *realistic* urban
rainfall-runoff model, and that benefits may be obtained from
a more detailed above-ground simulation on large catchments
as well as small. The viewpoint that it does not matter what
you do above-ground for routing in a mathematical model
appears to be ill-founded.

CONCLUSIONS AND FURTHER WORK

It is concluded that a realistic approach to the runoff phase
of urban drainage, such as the non-linear reservoir model
used in this investigation, will give superior results to
the time of entry approach of conventional storm drainage
design methods (Figure 2). Indications, as yet not checked,
are that this superiority extends to large catchments as well
as to very small ones. The time of entry approach leads to
errors both in timing and magnitude of inflow to the pipe
network. The extension of the experiment to a third adjacent
catchment has recently been completed. This is much larger
(approximately 10 ha), and already contains two gully meters.
The approach described in this paper will be tried on this
larger catchment. There remains the need to increase understanding of the physical behaviour of urban catchments. This
seems to be the best hope for improvement in prediction of
total runoff volume.

ACKNOWLEDGEMENTS

This paper is presented by kind permission of Dr. J.S.G. McCulloch, Director of the Institute of Hydrology, whose interest in the work at Southampton is acknowledged.

15

URBANIZATION AND NATURAL STREAM CHANNEL MORPHOLOGY: THE CASE OF TWO ENGLISH NEW TOWNS

C. Knight

Department of Geography, University of Exeter

ABSTRACT

The hydrological data for small catchments, encompassing Skelmersdale, Lancashire and Stevenage, Hertfordshire, indicate increases in the number and magnitude of flood-peaks, increases in total annual flows, and decreases in the seasonality of response to rainfall since the commencement of town development 15 and 31 years ago respectively. Increases in the mean annual flood appear to be of similar orders of magnitude for the two catchments, and the changes appear unrelated to climatic variation.

Long term changes in the sediment regimes of the previously rural catchments may also be expected, although the direction and order of magnitude of these may only be hypothesised.

Comparison of bankfull channel capacities for the urban streams with those of neighbouring rural streams in geologically and topographically similar catchments indicates increases of channel capacity of up to six times, and increased variability of capacity in, and below, the urban area. Larger increases are evident below Skelmersdale than Stevenage, and the effect is most marked immediately below the urban area, generally decreasing in significance downstream. The larger increases below Skelmersdale may be the result of channel adjustment to a flood of shorter return period, which has undergone a greater increase with urbanization than the longer return period floods to which the Stevenage Brook is adjusted.

The inferred changes in channel size are supported by evidence from surveys undertaken prior to urbanization, and by currently observable active undercutting and collapse of channel banks. This has implications for stream bank structures such as culverts and bridges below urbanizing areas.

INTRODUCTION

Hydrologists have recognised the significance of the rapid
growth in urban areas, and a relatively extensive corpus
of literature now exists outlining the extreme changes in
catchment hydrology which may be associated with paving,
sewerage systems, and other aspects of the built environment (Gregory, 1974; Hollis, 1975). The effects of construction activities in increasing the sediment yields of
catchments are also well documented in several areas
(Walling, Chapter 12; Guy, 1970) although the
associated long term effects of 'stable' urban areas are
less well understood.

The morphology of channels in erodible materials is
adjusted to the prevailing hydrological and sedimentological regime. The implications for channel form, of
increases in numbers and magnitudes of floodpeaks and of
changes in sediment loads have been outlined in a steadily
increasing number of papers, mostly concerned with the
eastern United States. British work has appeared recently
as the frequently detrimental effects of progressive
channel changes on the amenity value of rivers have become
apparent. The British work has generally followed the
earlier American ideas in methodology. Three main
approaches to the problem are evident, and these together
with a brief summary of the findings will be considered.

Most obviously, a series of direct measurements of a
channel in an urbanising catchment at intervals over an
extended period of time, such as outlined by Leopold
(1973) and Emmett (1974), can be used to assess the channel
changes consequent upon urbanization. In Watt's Branch,
a small tributary of the Potomac River in Maryland, twenty
years of such measurements of monumented cross-sections
showed an initial infilling of the channel with sediment
derived from construction sites. Following this a slow
enlargement occurred as the number of overbank floods
increased. However by the end of the study period the
channel was still some 20% smaller than its original size,
although further enlargement was expected.

From work in the eastern United States, Wolman and
Schick (1967) suggested that the time required for the
removal of infill was less than seven years. This agreed
with the earlier work by Guy (1965) who documented
complete removal of infilling sediments between 1957 and
1962 in a small catchment near Kensington, Maryland.
Wolman (1967) also indicated that, dependent on the
succeeding sequence of floodflows, channel enlargement
could be complicated by stabilisation of deposits with
vegetation or urban debris. Additionally, Dawdy (1967)
pointed out that the infilling deposits were frequently
coarse, and bank erosion either side of a central bar
might be precipitated.

The extended periods of time over which measurements
are needed to establish accurately long term trends
following completion of construction frequently make direct
monitoring impossible, or at best inconvenient. Furthermore, the setting up of monumented sections on British
streams which are liable to 'improvement' during flood

prevention schemes may prove hazardous. Consequently, no results of this type of study in Britain are known.

Several studies have used various types of survey material gathered at earlier dates for other purposes, in order to document channel changes. Graf (1975) used air photographs to show the extension of floodplains associated with the construction phase of urbanization in Denver, Colorado. Similarly, Hollis and Luckett (1976) used a survey of a short section of Canon's Brook below Harlow, Essex, undertaken by Harlow Development Corporation in 1956, and compared this with a partial resurvey made in 1970. This appeared to show a greater frequency of erosion than deposition at the nineteen sections examined. However the result was statistically insignificant possibly because of the continuing high sediment loads in the stream. Problems in the relocation of sections may also have led to some error.

The River Bollin in Cheshire was the subject of a further study of this type by Mosley (1975). Early large scale Ordnance Survey maps and more recent air photographs were examined to investigate channel changes in a catchment which has had a steadily increasing population since the early 1930s. From 1872 to 1935 the channel appeared to be in quasi-equilibrium, but a dramatic increase in the rate of channel shifting and a decrease in sinuousity occurred after 1935. Significant increases in channel width at various sections were also found. As no progressive variation could be identified in a number of climatic parameters, and significant increases in the maximum annual floodpeaks since 1956 were noted, the changes were ascribed to progressive urbanization. Further support for this hypothesis was provided by the coincidence in time of the geomorphic, hydrologic and land use changes.

The third method of assessing actual long term channel cross-sectional changes relies on 'prediction' methods for estimating the channel size prior to urbanization from channels in rural basins. Change is inferred by comparing this with existing channels in, and below, urban areas. A common frequency of bankfull flood flows for the whole channel or area is assumed. A further space-time or ergodic substitution can also be made, with areas of various degrees of development representing stages in the progressive urbanization of a basin.

Among the first to use this method was Hammer (1973), for small basins on the outskirts of Philadelphia, in the northeastern United States, but the method has also been used in Britain. Basically, a regression equation is derived relating measured bankfull channel cross-sectional area, or capacity, to drainage area (and possibly to other variables such as channel slope) in rural catchments either adjacent to or upstream from the urban area. Both methods of deriving such a 'rural' relationship may be criticised, the former on the grounds that 'space-time substitution is only really applicable downstream along individual channels' (Park, 1977); and the latter because of the sometimes inacceptable extrapolation of relationships far outside the limits of the data, for prediction purposes.

Hammer's first study (1973) demonstrated that for sections in quasi-equilibrium the degree of enlargement found was related to the type of urban land use in the catchment. The degree of enlargement varied from zero to seven times the original channel size, impermeable areas, sewered streets and housing areas having greater effects than other urban areas such as new houses, golf courses and houses older than thirty years. However, it was noted that the high level of intercorrelation between various aspects of urbanization probably allowed some independent variables to 'borrow' causal influences that should properly have been attributed to other factors. Natural watershed features such as slope, basin size and the location of building within a catchment had a minor impact on the enlargement ratio.

The British work undertaken using similar methodology has generally shown changes in channel capacity of the same order of magnitude as those found in the Philadelphia region. However, in many cases the proportion of the catchment under urbanization has been low, and in the case of Holsworthy, on the River Deer, and Woodbury, also in Devon (Park, 1976), the built-up areas are not served by a separate surface water sewer system. Average enlargement ratios of 1.44 below Holsworthy, 2.61 below Woodbury, 2.62 below Catterick Garrison in Yorkshire (Gregory and Park, 1976a) and 5.31 below road drains in central Devon (Gregory and Park, 1976b) are reported, whilst the work by Hollis and Luckett (1976) in West Sussex suggested that a 10% paving of the catchment would result in an increase in channel size of 1.7 times.

These British cases include examples of many types and degrees of hydrological modifications. These include the effects of direct surface water drainage, transferred water inputs, sewage effluent inputs from foulwater and combined systems after treatment, overflows of untreated sewage from combined systems, and the effects of varying sizes and locations of urban areas within the catchments. Consequently, any subsequent interpretation of the results is difficult to undertake as the degree of hydrological change in each case in unknown. As the sediment loads of most British rivers are low, on a world scale, it is probably unreasonable to infer channel modifications unless substantial changes in peakflows can be hypothesised. It has also been pointed out (Richards and Wood, 1976) that changes in channel shape, roughness and velocity are also likely to occur, and that the amount of cross-sectional enlargement to be anticipated might be decreased as a result.

To overcome these difficulties, studies of channel modifications are needed in streams where the degree of hydrological modification is known, or can be deduced, and for which the urban area occupies a large proportion of the catchment. This precludes the use of extrapolation techniques based on relationships established in upstream rural reaches, as in catchments with a large proportion of urbanization only a small area will be available to develop this relationship.

THE STUDY AREAS

The New Towns of Britain provide potentially interesting
cases for investigation because growth of the urban areas
and their associated surface water sewerage systems are
well documented, and large scale plans of various types
for different stages of development are often available.
Furthermore, artificial modifications to the stream
channels, which are very common and frequently undocumented
in older urban areas, can be identified and isolated. In
this study, questionnaire surveys completed by New Town
Development Corporations were used to select areas for
examination. The two areas finally selected for analysis
were of similar size, and both encompassed large areas of
urbanization; but they were strongly contrasted in terms
of geology, and age of development. In both areas the
foulwater drainage system made no impact on the hydrological regime, and hydrological records suitable for a
partial analysis were available for part of the period of
urbanization.
     Stevenage, in Hertfordshire, lies in rolling country
on the dip slope of the Chiltern Hills, within the catchment of the Stevenage Brook. The area is underlain by
Upper Cretaceous Chalk, with up to 30 m of overlying
deposits of boulder clay, clay-with-flints, glacial sands
and gravels, and alluvium, whose configuration has given
rise to much speculation on the glacial history of the
area (Clayton and Brown, 1958). The mapping of these
deposits, undertaken by the Geological Survey in 1860, and
partially resurveyed by the Soil Survey between 1947 and
1949, may be regarded as somewhat tentative, owing to the
great variability in lithology, and to identification
problems. Stevenage Brook, and its tributary Aston Brook,
are incised in alluvial deposits for most of their courses.
The mean annual rainfall of the area is some 620 mm at
Broomin Green, increasing to around 650 mm at 120 m O.D.
in the north of the catchment. The urban area of
Stevenage lies almost entirely within the catchment of
the Brook, with the exception of a very small area
in the northeast, which drains by way of the Walkern
valley to the River Beane.
     The 1976 population was about 70 000, and this has
grown steadily from a population of about 6000 since 1946
(see Table 1). Analysis of the area covered by urban land
use at successive dates, based on Development Corporation
records, shows a fairly constant growth rate. Slightly
faster rates occurred between 1955 and 1958 when housing
densities were lower than at present, and when the main
commercial centre was under construction. Growth after
1973 has slackened. The total area under urban land uses
of all types represents 55% of the drainage area of the
Bragbury Park gauging station on the Brook. The main
lines of the very extensive surface water sewerage system
were laid down at an early date, central areas being
served before 1953, the majority of the main industrial
area by 1955, and most other areas by the late 1950s
(see Figure 1). This has resulted in a very large increase
in the effective drainage density of the catchment

Table 1.   Comparison of some aspects of the studied
catchments.

	Stevenage	Skelmersdale
Date of designation of New Town	1947	1961
Initial Population of catchment	6000	10000
Population of catchment in 1976	76000	39000
Area of gauged catchment (km²)	32.11	20.80
% of catchment under urban land use	54.9	59.7
Natural drainage density from 1:25000 Ordnance Survey maps	0.589	1.155
Effective drainage density post urbanization including main surface water sewers	6.970	2.030
Period of hydrological records	1958-1976	1967-1973

from the natural Ordnance Survey 1:25000 'blue line network'
value of 0.59 km km^{-2}, to 6.97 km km^{-2} when the main sewer
lines are included (see Table 1).   Small on-channel flood
balancing ponds have also been constructed at various points
along the channel, but these appear to have had little
effect on the outflow hydrographs at the gauge.   Skelmersdale, Lancashire, presents a broadly similar outline in
terms of relatively uniform growth of the developed urban
area, although the area was designated at a later date, in
1961, and had a higher initial population of some 10000,
mainly resident in the villages of Skelmersdale and
Upholland.   The area was predominantly rural prior to 1961,
but by 1976 the population had reached 39000 and 62% of the
catchment was under urban land use.   The urban area drains
into the River Tawd, although a small area of houses and
gardens built prior to 1961 drains into the river below the
main gauging station at Cobbs' Brow.   As is frequently
the case in British towns, extensive culverting of
the pre-existing natural tributaries has been undertaken
(Figure 1), and this includes a stretch of the main channel
through the town centre.   This has resulted in a more
modest increase in the effective drainage density, from
1.16 to 2.03 km km^{-2}.

The catchment of the Tawd is mainly low lying and is
underlain by heavily faulted Middle and Lower Coal Measures,
with small areas of Millstone Grit, Bunter Sandstone and
Manchester Marl.   As in the case of Stevenage Brook, considerable depths of superficial drift are found, including
boulder clay, Shirdley Hill sands and a small area of peat.
These deposits are undoubtedly of significance in influencing the hydrological regime of the catchment, as has been
shown in neighbouring areas by Worthington (1977).   The
channel of the Tawd is cut mainly in alluvial deposits,
although occasional outcrops of bedrock appear in the
channel bed and banks, restricting the ability of the
channel to enlarge in one or more dimensions.   The area

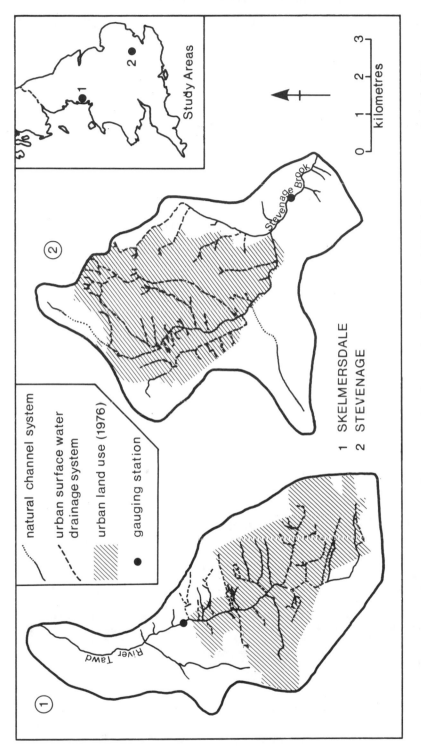

Figure 1. The urban study catchments, showing natural and artificial components of the drainage systems, and areas under urban land use in 1976.

receives some 870 mm of rainfall per annum, mainly during
the winter months.

HYDROLOGICAL CHANGES

For both areas an investigation of rainfall characteristics
for stations within the catchments revealed no significant
trends. Although changes in other climatic parameters
are possible it is unlikely that they would account for
the magnitude of the streamflow changes described, and
consequently no further investigation was performed.

Both catchments have gauging stations below the pres-
ently built-up urban areas, but as mentioned previously,
the records are inadequate for a complete assessment of
hydrological trends within the catchments, and could only
be used to give an idea of the direction and order of
magnitude of any changes. The gauge below Stevenage is
sited in Bragbury Park (see Figure 1). Several alterations have been made to the gauging structure since
records began. It now consists of a broad-crested weir
with chart recorder, plotting discharge directly through
appropriate gearing and chart scales. Prior to this a
Munro stage recorder was in operation for a short period,
and the earliest recorder, on a 'Venturi' flume, was of a
type more usually used for gauging sewage outfall discharges. The earliest charts held by Stevenage Development
Corporation date from September, 1958, when approximately
21% of the catchment was under urban land use, although
there is evidence to suggest that the recorder was in use
before this, and that the records have since been lost.
Daily charts were in use for most of the period, and these
had frequently been left unchanged for several days, making
the trace difficult to follow. Incorrect placing of the
chart on the recorder drum is also found, causing errors
in the representation of peakflows, and there are many
occasions on which other malfunctions appeared.

The records from Cobb's Brow, on the Tawd below
Skelmersdale, date from October 1967, when some 17% of the
catchment was under urban land use. The records continue
until October 1975, but from March 1973 to November 1974
the gauge was inoperable owing to valdalism, a recurrent
hazard for instrumentation in urban areas. Gauging was by
a weir and rating curve, and Munro stage recorder.

Recent geomorphological research has suggested that
bankfull channel capacity is adjusted to bankfull flows.
The frequency of this flow, as Harvey (1969) has pointed
out, may vary from stream to stream, and along individual
streams with complex regimes. However, it usually recurrs
approximately every one or two years. Consequently
hydrological analysis was restricted mainly to a study of
flood flows.

Stevenage Brook

The annual frequency of occurrence of floods in Stevenage
Brook has shown significant trends over the last seventeen

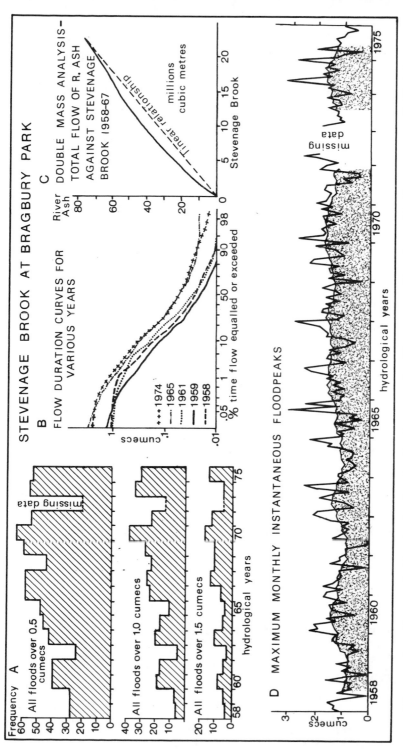

Figure 2. Some of the hydrological changes in Stevenage Brook. A) Frequency of floods above selected thresholds. B) Changes in flow duration curves for various years. C) Non-linear relationship of cumulative total flows in Stevenage Brook and the River Ash. D) Maximum monthly instantaneous floodpeaks, with 13 month moving mean.

years of record, both at relatively low thresholds
(0.5 m³/s) and at higher ones (1.0 and 1.5 m³/s). At
higher thresholds still, the small number of floods recorded make it difficult to isolate a definite trend
(Figure 2). For all floods over 1.0 m³/s there appears to
have been an approximate trebling in the frequency of
occurrence between 1959 and 1975, from about 10 events per
year to thirty, a trend significant at the 1% level.
Analysis of the proportional distribution of floodpeak magnitudes in successive periods indicates that the proportion of floods between 1.0 and 1.25 m³/s has increased,
whilst the proportion between 0.5 and 1.0 m³/s has fallen,
despite an increase in the absolute frequency of occurrence.
Higher increases in the frequency of medium sized floods
than in lower ones were also noted by Hollis (1974) in his
study of the influence of urban development on flows in the
Canon's Brook, Essex. The increased frequencies appear to
have affected winter (October to March) floods slightly
more than summer floods, but this effect is not marked.
    The magnitude of floodpeaks also appears to have been
affected by the urbanization, and moving means through the
maximum monthly instantaneous peaks show an increase since
1958. The annual maxima show a similar trend, giving an
increase of about 50% over the period, but this is only
significant at the 5% level, all other trends being significant at the 1% level. As the data are clearly a non-
stationary series, and no record of pre-urbanization flows
is held, a consideration of partial duration series is
strictly questionable. However, a tentative investigation
reveals that an increase of about 40% in the mean annual
flood has occurred between the first and last quarters of
the period of record, with slightly larger increases in
floods of shorter return periods. Assuming that this
increase was proportional to the increase in the proportion
of the catchment occupied by urban land use would suggest
an increase of 243% in the mean annual flood over the
whole period of construction, from 1.15 to 2.8 m³/s. This
is very similar to the 250% increase suggested by Hollis's
(1975) analysis of the literature for the 18% paving of the
catchment.
    In addition to the changes in flood magnitudes, flow
duration curves for selected years, based on three hourly
readings, indicate that increases were associated with
almost the entire range of flows, including flows of less
than 0.02 m³/s. Double mass analysis of total flows in
the Brook, against total rainfall received at Broomin
Green, and against the total flow in the River Ash, a
nearby rural stream, provide further support for these
progressive increases in streamflow over the whole period
of record.

River Tawd

The analysis of the hydrological data for the Tawd below
Skelmersdale revealed similar trends to those found in the
Stevenage Brook, although the results must necessarily be
more tentative because of the shorter period of record

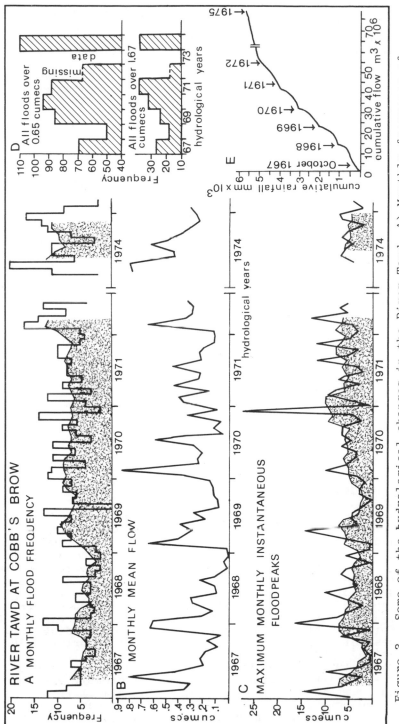

Figure 3. Some of the hydrological changes in the River Tawd. A) Monthly frequency of floods greater than 0.65 cumecs, with five month moving mean. B) Monthly mean flow. C) Maximum monthly instantaneous floodpeaks with five month moving mean. D) Annual frequency of floods above selected thresholds. E) Total cumulative flow in the Tawd, against total cumulative rainfall received at Cobb's Brow gauge.

(see Figure 3). Increases in the frequency of occurrence of floods greater than 0.65 and 1.67 m^3/s can be picked out from the five month moving means, but the annual total of floods greater than 0.65 m^3/s shows an increase only significant at the 5% level, probably as a result of the shortness of record. There is less evidence than from the Stevenage Brook to suggest proportionally larger increases in numbers of medium sized floods compared with smaller ones.

A five month moving mean through the maximum monthly instantaneous floodpeaks indicates an overall increase throughout the period until the end of 1972, although after the break in the data this increase is not evident, probably as a result of the very dry weather experienced throughout the majority of this latter period. Partial duration flood frequency plots of the first and second half of the period of record indicate an approximate increase of 18% in the mean annual floods, from about 17 to 20 m^3/s. By extrapolation, this would suggest a mean annual flood of 12 m^3/s prior to development and 22.2 m^3/s by 1976. This increase is smaller but of a similar order of magnitude to the change estimated by Hollis (1975), of 260% for a 20% paving of the catchment. It is a little less than the change in Stevenage Brook flows, probably as a result of the great geological and climatic differences between the catchments.

Mean monthly flows also appear to have increased, especially in the summer months (March to September). This had greatly reduced the seasonality of flows by the summer of 1972. The loss in seasonality, and the increase in total flows is evident in double mass analysis of runoff against rainfall at Cobb's Brow gauge. Average summer runoff was some 15% of rainfall at the start of the period, but increased to around 45% by 1972. Total winter runoff was less affected, remaining at around 50% of rainfall throughout the period.

CHANNEL CHANGES

In a short term research project, measurement of long term progressive changes in a stream channel cannot be undertaken, and reliance must be placed on other methods to estimate any channel modifications. Some evidence of current channel change is given by the presence of very extensive undercutting of structures such as bridges, and the presence of large numbers of collapsed trees on slumped and undercut banks. Such evidence is subjective however, and not necessarily indicative of channel enlargement. For the most part analysis must rest on prediction techniques for estimation of channel dimensions prior to urbanization.

As Park (1977) has pointed out, streams in the same area may show marked variations in channel capacity, and consequently it is important to ensure comparability in terms of geology, climate and topography, between catchments used to calibrate the prediction model, and the studied urban catchment. Climatic homogeneity and

similarity in solid geology are relatively straightforward
to achieve, but superficial deposits such as glacial drift
are frequently very varied, and poorly mapped.   This
causes problems in maintaining comparability between catchments, problems which are also evident in attempts to
ensure similarity in topographic and morphometric parameters such as drainage density, basin order, mainstream
slope, relief ratio, and basin shape.   Inevitably, com-
promises have to be made in order to evolve any generalised
rural relationships, and in the absence of a very large
data set, this analysis was restricted to selecting the two
rural catchments most similar in all respects to the study
basins.

Ippollitts Brook and the Old Bourne catchments, in
Hertfordshire, were selected for comparison with Stevenage
Brook, and Rainford and Eller Brooks, in Lancashire, for
comparison with the Tawd.   The major differences between
the Hertfordshire streams lay in relief and drainage
density factors, the rural streams having somewhat steeper
catchments with greater drainage densities.   Logically,
however, one might expect this to lead to larger channels
in the rural catchments, as runoff will be more rapid, and
consequently any channel enlargement found will be a conservative estimate.   In the case of the Lancashire streams
the main differences were again in the relief parameters,
with the catchment of the Tawd being steeper than the two
rural ones, although the difference in mainstream slope
was less marked than the difference in overall catchment
relief, and other parameters showed very close similarity.

Fieldwork in the two areas was carried out during
1976, when data on channel capacity were collected at 123
sites in Hertfordshire, and 91 sites in Lancashire.
Bankfull channel capacity was identified in the field.
This identification was based on breaks in the lope of
channel walls, on changes in the vegetation on the sides
of the channel and adjacent berms, and on the position of
flood debris and recently deposited sediment.   As a result
of incision, the upper surface of the 'bankfull channel is
frequently not coincident with the level at which extensive
flooding begins to occur' (Gregory, 1976).   Following this,
the relationship of bankfull channel cross-sectional area
to drainage area, measured by Precision Disc Planimeter
from 1:25000 Ordnance Survey maps, was established by
regression analysis.

In the case of the Hertfordshire streams, the relationship between channel capacity (C) in m^2 and contributing
area (A) in km^2 for the rural catchments was of the form

$$C = 0.307 \, A^{0.347}$$

The relationship is shown in Figure 4, with confidence
limits of one standard error about the regression line
marked.   Tests for differences between separate relationships for the two rural catchments showed no significant
differences at the 1% level.   The relationships between
width and depth and drainage area are also shown, but they
are less strong, probably as a result of the effects of
varying channel materials on channel shape.   For these

193

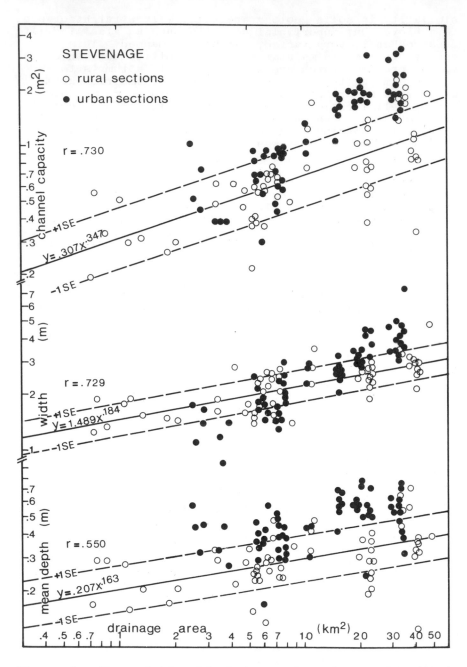

Figure 4. The relationship between channel capacity (top),
width (centre), and depth (bottom) and drainage area,
for the Hertfordshire channels. The regression lines
and confidence bands refer to the rural channels.

streams the relationship between contributing channel
length and channel capacity was actually stronger than the
relationship shown (r = 0.81 as against r = 0.73), probably
because of the improved representation of areas actually
contributing to peakflow in this area of low drainage
density. However the resulting differences in channel
capacity at any point were small, and consequently the use
of drainage area was maintained, for comparison with the
Lancashire data.

For the rural Lancashire streams, a strong relationship between channel capacity and drainage area, of the
form

$$C = 0.506 \ A^{0.436}$$

was evident (Figure 5). The reduced scatter about the
line, in comparison with the Hertfordshire data, is under-
standable in that the catchment is underlain by a less
variable and permeable medium. Use of contributing
natural channel length as the independent variable resul
ted in no significant improvement in the level of explanation, and was therefore not considered in the remainder
of the analysis. Comparison of these relationships with
those for the data for the urban channels indicate marked
differences in channel sizes, especially in the reaches
immediately below the urban area where the proportion of
the catchment occupied by urban land use is large. In the
case of Stevenage Brook, the majority of the sections
lying immediately below the urban area lie more than one
standard error, and several lie more than two standard
errors, above the calculated regression line for the rural
channels. Larger increases appear to be evident in the
depth dimension than in width, but this conclusion must be
more speculative, for reasons outlined above.

The inferred enlargement is greater and more significant in the Tawd below Skelmersdale, again particularly
immediately below the urban area. Part of the increased
capacity in some sections is undoubtedly the result of
localised scour immediately below surface water sewer
inputs to the channel, but the rest is attributable to the
changed hydrological, and possibly sedimentological,
regimes. In this case the increases are more apparent in
the width dimension than in depth, possibly as a result of
the occasional outcrops of bedrock in the channel floor,
restricting changes in the depth.

The calculated enlargement ratio, or the ratio of the
predicted to the actual channel capacity, varies from
about one at the upper end of the Tawd, to between three
and four immediately below the urban area, although a ratio
of six is found below the section of culverted mainstream.
As the proportion of the catchment occupied by the urban
area decreases, moving further downstream, the ratio drops
to around two. This may be the result of proportionally
smaller changes in catchment hydrology at this point.
Alternatively, the channel may have had insufficient time
to adjust to the changed regime, due to the increased
amount of sediment to be removed for a similar proportional
increase in channel size. In the case of Stevenage Brook,

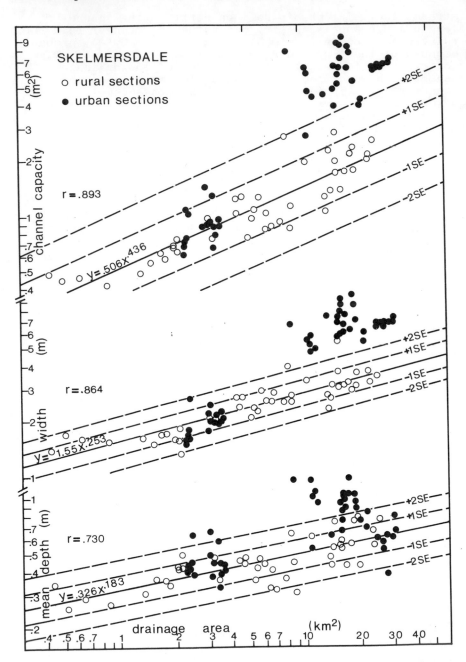

Figure 5.   The relationship between channel capacity (top),
      width (centre) and depth (bottom) and drainage area,
      for the Lancashire channels.   The regressions lines
      and confidence bands refer to the rural channels.

there is a steadier increase in the enlargement ratio
through the urban area, from about 1.5 at the upper end of
the natural channel within the urban area, to 2.3 at the
lower end, although again there are isolated higher ratios
below surface water sewer inputs. The decrease in the
ratios to about 2.2, moving away downstream to the confluence with the River Beane, is less marked than in the
Tawd.

Some confirmation of these findings may be made by
referring to plans available for the two areas prior to
the development of the New Towns. For the Stevenage area,
plans of a short stretch of the Brook within the urban
area were made early in the history of development,
although the exact date is not known. The survey consists
of a plan, and 26 cross-sections of the channel at a scale
of 1:120. Identification and analysis of the bankfull
channel capacity in each section shows a scatter of points
about the rural regression relationship. The scatter in
the data is quite large owing to the difficulty in iden-
tifying bankfull channel capacity in this incised section,
from the plans. In addition, the reach is not one in
which the largest amount of channel enlargement appears to
have occurred, and it lies a considerable distance above
the gauging station, and hence the degree of hydrological
change is unknown.

In the case of Skelmersdale, plans at a scale of 1:500,
based on a photogrammetric survey carried out in 1961,
were used to analyse channel widths prior to the start of
construction activities for a channel reach in which the
apparent channel enlargement was very large. Water
width, treeline width and areas of scour were marked, and
an average taken for each of 37 cross-sections. The
version of Student's 't' test used by Hollis and Luckett
(1976) for a similar analysis, and suitable for dependent
samples made up of matched pairs, was used to test for
differences in channel widths between sites examined in
the field in 1976, and the same sites measured from 1961
plans. The average increase in channel widths was 1.72 m
which was significant at 0.1%. A further analysis
including 24 more channel widths taken from the plans,
showed a similar relationship to drainage area as the
rural channels, giving further support to the idea of
channel change since 1961.

The differing amounts of channel enlargement in the two
areas are difficult to explain purely in terms of the
relative magnitude of hydrological changes. However, some
clue as to the cause of the greater enlargement in the Tawd
may be provided by an examination of the bankfull channel
capacities of the two streams. In surveyed sections of the
Tawd near the gauging station, the application of Manning's
equation reveals that the channel is adjusted to a flood
with a recurrence interval of between 10 and 14 months
(.83 to 1.67 years). This is in contrast to the Stevenage
Brook, with a less flashy regime, which appears adjusted
to a flood with the longer recurrence interval of about 27
months (2.25 years) at present hydrological conditions.
Work by Harvey (1969) similarly suggested longer return
period floods as the controlling influences on channel size

in less flashy chalk streams.

Hollis (1975) and others have pointed out that the floods of shorter return periods are more affected by urbanization than larger ones, as during storms of long return period the catchment evidences a 'saturated' response whether under urban land use or not. It is highly likely that this is the case in the Stevenage Brook and Tawd catchments, although the poor quality of the hydrological data make this impossible to ascertain accurately. The greater enlargement in the Tawd than in Stevenage Brook may therefore be the result of proportionally greater increases in the bankfull, channel forming flood. Changes in channel slope, roughness and shape are also probable, and the fuller explanation of all aspects of the system awaits more detailed investigation.

CONCLUSION

Substantial changes in the hydrological, and probably the sedimentological, regime of two catchments which have been progressively urbanised, have been demonstrated to lead to changes in channel morphology. These channel changes have been investigated both by indirect means, and by examination of historical material. Other areas of Britain have shown hydrological changes as great as those noted in this study, and the effects of channel enlargement in such situations cannot be ignored. The impact of increases in width and depth of natural channels on stream bank structures such as bridges and culverts below urbanising areas is readily apparent. Planning strategies should bear this, as well as the perhaps more obvious hydrological impact, in mind.

ACKNOWLEDGEMENTS

The author acknowledges the assistance of Stevenage and Skelmersdale Development Corporations in making available substantial amounts of data. This research was carried out with the financial support of the Natural Environment Research Council, and thanks are due to Professor K.J. Gregory, Dr. D.E. Walling and Professor A. Straw for help and encouragement.

16

THE NATURE AND SOURCES OF URBAN SEDIMENTS AND THEIR RELATION TO WATER QUALITY: A CASE STUDY FROM NORTH-WEST LONDON

J.B. Ellis

Middlesex Polytechnic, Hendon, London

ABSTRACT

The pollution loads resulting from stormwater discharges to urban streams are primarily exerted by high concentrations of particulate materials. The character, nature and sources of urban sediments discharged through a separately sewered development in the north west suburbs of Greater London are examined and their relations to and significance for the quality of receiving waters is demonstrated. The effectiveness of current street cleaning practices is questioned and the prime importance for water quality of the organic and solids fractions below 0.06 mm is established.

INTRODUCTION

Only in the last twenty years have attempts been made to quantify the levels of pollution from urban stormwater discharges which contain substantial quantities of particulate materials.
    The U.S. Federal Water Pollution Control Federation (1969), AVCO Economic Systems Corporation (1970) and Whipple et al., (1974) have demonstrated that the quality of urban runoff is normally no better than secondary sewage treatment effluents with the yields of particulate material often being well above raw sewage quality. Studies in Cincinatti (Weibel et al., 1964) showed that storm discharges from urban surfaces contained 1.4 times as much suspended solids as raw sewage. Bryan (1970) in Durham, North Carolina, recorded suspended solids yields 20 times that contained in municipal wastes and Colston (1971) showed that even if Durham provided 100% removal of organics and suspended solids from the raw municipal waste on an annual basis, the total reduction in solids discharged to receiving waters would only be some 5%. The mean

suspended solids concentrations of urban stormwater observed by Cordery (1977) in Sydney, Australia was equal to that of local raw sanitary sewage, although yields 5 times that of raw sewage were often recorded.

The first-flush peak of suspended solids during storm events in urban areas can also attain exceedingly high levels. Palmer (1950) recorded total solids concentrations of 914 mg/l for storm flows over paved surfaces in central Detroit although he noted marked variation in concentrations between points and at the same point during runoff. There was considerable spatial and temporal variation in solids yields, in some cases the quality deteriorating as the storm progressed, in others becoming better, whilst other points showed no apparent pattern. Wilkinson (1956) recorded suspended solids yields of up to 2045 mg/l for stormwater discharges from a 247 ha. housing estate at Oxhey, Hertfordshire, and commented on the sudden variation in solids concentration that can occur during any specific storm event. Over a quarter of all samples recorded suspended solids concentrations greater than that of sewage effluents from the housing estate. Waller (1972) in Halifax, Nova Scotia, although reporting low values of suspended solids for stormwater runoff, averaging between 59 to 132 mg/l with peak values of 645 mg/l, was able to show that such concentrations equalled or exceeded that of the combined sewage system serving the catchment for at least 30% of the time. The Water Research Centre investigations of a separately sewered 136 ha. residential catchment at Shephall, Stevenage, have recorded suspended solids concentrations of up to 2300 mg/l (G. Mance, personal communication). It is estimated that the existing separate storm sewer system discharges a mass of pollutant equivalent to that which would be dishcarged from an overflow set at three dry-weather flows on a hypothetical combined sewer system for the same catchment'.

The drowning and containment of combined sewer overflows, through increase in stage of the receiving stream during storm flow, means that highly polluting slugs of solids material can be released to the river as the flood level subsides. Under such conditions the maximum dilution capacity of the flood peak flow is not available and major pollution loads occur (Appleton, 1975). De Filippi and Shih (1971) noted that for short duration storms in combined sewer discharges in Washington, D.C., waste concentrations increase with discharge rate, with the peak concentrations concurrent with peak flow rate. For separated systems, average organic concentrations were one third of those for combined sewers, although solids concentration in the separate system exceeded that of the combined system by a ratio of 4 to 1.

The effects of the injection of massive doses of particulate matter to receiving streams in urbanised areas would thus suggest that urban stormwater runoff is normally no better than secondary sewage treatment effluent and is often very much worse. The current trend of national policies and abatement programmes for the elimination of pollution from point discharges (Whipple, 1975; Lowing, 1976; Haith, 1976) implies that greater priority

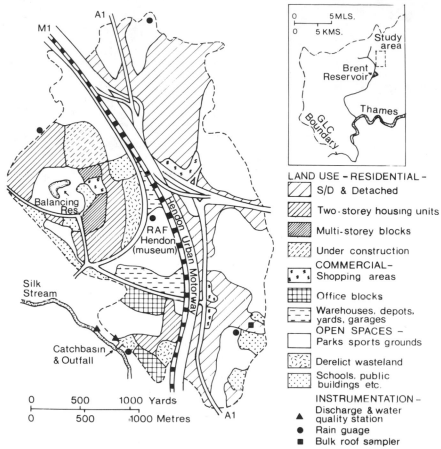

LAND USE - RESIDENTIAL -

◻ S/D & Detached

◻ Two-storey housing units

◻ Multi-storey blocks

◻ Under construction

COMMERCIAL-

◻ Shopping areas

◻ Office blocks

◻ Warehouses, depots, yards, garages

OPEN SPACES -

◻ Parks sports grounds

◻ Derelict wasteland

◻ Schools, public buildings etc.

INSTRUMENTATION -

▲ Discharge & water quality station

● Rain guage

■ Bulk roof sampler

Figure 1.   The Graham Park catchment: land use and instrumentation.

will be given in the future to pollution from combined sewers, separate storm drains and overland flow. The contribution from separate stormwater sewer systems to the pollution of receiving waters by high solids loadings would appear to be significant. Adequate and prudent water resource planning would therefore dictate that any urbanised area which has both types of sewer system must consider and evaluate the entire storm flow problem before proper treatment and upgrading programmes can be devised and implemented. McPherson (1969) has stated that if stormwater discharges are found to require treatment, then pollution abatement of combined sewer overflows cannot be mitigated by conventional separation of combined systems into separate sanitary and stormwater systems. Consequently, urban area planning to upgrade secondary sewage treatment plant, because of possible contravention of stream quality standards, need to carefully assess the pollutional potential or urban stormwater discharge.

THE STUDY AREA

Stormwater discharges have been monitored for a 3.5 km^2

201

Figure 2.   A general view of the Graham Park catchment.

catchment on the site of the former Hendon aerodrome in
north-west London.   With an overall housing density of
15 units per hectare and an estimated population of 12000,
the Graham Park catchment is drained on a separate sewer
system to the Silk Stream, a tributary of the River Brent,
which outfalls to the Brent Reservoir (Figure 1).   The
surface water sewers were designed by the TRRL method on
5 year Bilham storm probabilities using a rainfall
intensity of 1.78 mm/hour and an assumed 35.5% impervious
area.   Land use within the catchment comprises 30.5% of
the area in residential multi-storey blocks (Figure 2),
open spaces occupy 26.1%, public and official grounds
(Ministry of Defence and Metropolitan Police property,
schools and the RAF Museum) occupy 22.7%, warehouses,
offices and shops cover 5.9% and light industrial premises
take up 1.9%.   Almost 10% of the area comprises highways,
including 0.16 ha of the Hendon Urban Motorway.

INSTRUMENTATION AND METHODS

The outfall to the stormwater system is by a 3.65 m x 1.22 m
concrete box culvert into a catch basin sunk below the level
of the receiving stream (Figure 1).   Water level at the
culvert outfall is monitored by an Arkon Instruments air-
purged system, with the dip tube in the culvert connected
to the recorder housed above the culvert.   A stage-discharge
relation for the system has been derived from formulae
supplemented and validated by subsequent current meter and
volumetric rating.   Water quality monitoring of storm events
by an automatic pumping sampler, housed in the recorder
station, provides a sampling rate of 325 ml/min and is
triggered by means of a suspended mercury float switch
activated by the rising limb of the storm wave.
        Intake line velocity can have an appreciable effect on
the sampled suspended sediment concentrations (Marsalek,
1976).   With line velocities ranging from 0.07 m/sec to
0.48 m/sec the sampler efficiency varied from 45 to 92% for
solids collection.   Therefore, the solids concentration of
the samples will always be too low, this being particularly

significant for particle sizes above 5 mm. As a check on this problem, the automatic sampler is calibrated over selected storm events by using a purpose built depth-integrating hand sampler (Ellis, 1975). The orientation of the automatic sampler intake is an additional source of error. Depending on the ratio of intake to flow velocities, the adoption of an orientation orthogonal to flow can under-estimate the sample concentration by 10 to 30%, although particle sizes below 75 microns are largely unaffected by flow ratios. Preliminary studies also suggest that during high flows there can be a considerable variation in concentration through the flow depth with mean concen-trations occurring at about the mid-point of the flow depth. This stratification breaks down with increases in turbulence, but the vertical variation complicates the calculation of pollutant mass transport.

Precipitation is determined from a network of rain-gauges (Figure 1) and a 3.2 litre tipping bucket instrument for roof runoff has also been installed following designs developed by the Water Research Centre, Stevenage. Its data logger will provide a continuous record of the timing as well as frequency of sampling to enable roof flow hydrographs and chemographs to be constructed.

HYDROLOGICAL CHARACTERISTICS AND WATER QUALITY

Storm precipitation contains particulate matter, gases, and photochemical oxidants purged from the urban atmos-phere. In nine storms the suspended solids concentration of rainwater varied between 0.5 to 13 mg/l with particles mostly below 40 microns and so rainfall contributed only 1 to 2% of the total solids loading of the stormwater runoff. Roof runoff increased the levels very considerably, with concentrations of 85 to 930 mg/l, equivalent to between 10 to 30% of the suspended solids loading to the outfall discharge. The peak concentrations seem dependent on length of preceding dry period and storm intensity, although local ground activities, such as building construction are also important influences.

The large impermeable area produces a considerable peaking of the outfall hydrographs (Figure 3) with rain-fall-runoff lag times of 30-40 minutes. Rising stormwater flows are accompanied by a marked first-flush phenomena with suspended, volatile and dissolved solids flushing out of the system. For a considerable time after the peak discharge runoff remains highly turbid as indicated by the suspended solids concentration on the recessional limb. This implies that the catchment area cannot be rapidly cleaned by the first flush of stormwater to the sewer system (Wilkinson, 1956). The first flush of particulate materials varies in concentration between 85 to 4500 mg/l with the relation of the solids peak to discharge peak being quite variable. Some storms show a very clear lag of the flood wave behind the sediment wave (storm of 18.10.76 in Figure 3). A tentative explanation of this

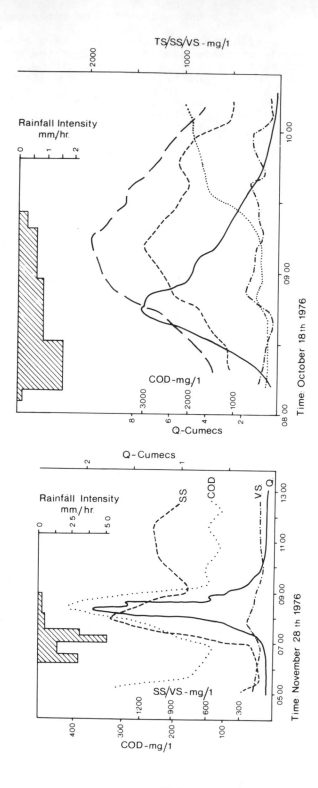

Figure 3. Hydrographs and chemographs for the Graham Park catchments.

leading prime sediment peak is the relatively poor flow
characteristics of the stormwater sewer system, particularly in the concreted culvert sections. Solids settle
and lodge in the system as lag deposits during antecedent
storm recessions but are rapidly and efficiently flushed
out on the rising limb of the storm wave and thus entrain-
ment precedes the sewer routing event often by as much as
0.4 hours. Fine solids and dust also settle and adhere
to expanding mats of sewage fungus and slime films during
low flow conditions between storm events. Large colonial
mats of *Sphaerotilus* fungus can accumulate rapidly and
their gelatinous filaments act as filters trapping large
amounts of fine particles and exacerbate the effects of
suspended and volatile solids on stormwater quality.
Runoff from road gullies very early in the storm event
loosens and breaks up these mats and provides another
explanation for the lead in the solids chemograph.

   The chemographs (Figure 3) typically show a double
peaking of suspended solids concentration. The subsidiary
peak on the recession limb of the flood represents fresh
sediment introduced to the street surface drains from
roofs, paths and driveways, and can be delayed behind the
prime peak by up to 3 hours depending on the intensity
and duration of the storm event. Stormwater concentrations of sediment normally fall after about 36 minutes
which is the time of concentration for the sewer system and
when the maximum dilution effects can be expected. The
chemographs for volatile solids show similar trends as
those for suspended solids indicating that substantial
amounts of rubber, bitumen, grease and oil are flushed to
the sewers, commonly as capsules of hydrocarbon coating
an inorganic nucleus. A notable feature of the sediment
discharge is the extreme range of concentrations that
result from storm events (Table 1). Variations in
concentration of one to two orders of magnitude are quite
common with the standard deviation normally being in the
range of two thirds to three quarters of the mean
concentration. The variability is greatest for materials
carried in suspension as particle sizes carried as a
washload or a suspended load can be transported by a
greater range of flows than can coarser solids which
require traction.

   An apparent linear trend exists between turbidity and
suspended solids concentration as seen in Figure 4. The
data scatter can be explained in terms of colour and
organic interferences, as well as by the effects of
flocculent micro-organisms on turbidimetric determination.
Adjustments for colour and organic matter increases the
correlation to 0.88, thus allowing 77.4% of the graphed
variance to be explained. Separation of the data points
on the basis of particle size and composition, as well as
by timing on the hydrograph, would indicate that turbidity
is highest during the early stages of flow on the rising
limb when organics and clay-silt sizes occur most
frequently. As peak flow is reached the sand-silt ratio
increases to 3:1, falling back to 2:1 on the recession
limb. The reduction in light penetration brought about by
the increase in turbidity levels, as well as the burial of

Table 1. Variations in selected parameters for storm
events.

PARAMETER	X̄	σ	Range	Coefficient of Variation
Total Solids (mg/l)	920.0	580.6	3150-90	0.631
Suspended Solids (mg/l)	546.5	426.3	2688-20	0.780
Total Volatile Solids (mg/l)	274.3	104.7	1605-28	0.382
Suspended Voltaile Solids (mg/l)	146.8	159.7	1268-8	1.088
BOD (mg/l)	22.2	13.8	672-3	0.622
COD (mg/l)	342.8	231.3	2105-31	0.675
pH	7.38	1.3	7.95-6.63	0.174
Alkalinity	174.1	151.7	806-20	0.871
Total Hardness	186.6	82.5	316-80	0.442
Total Coliforms	230.6	240.1	2407-2	1.041

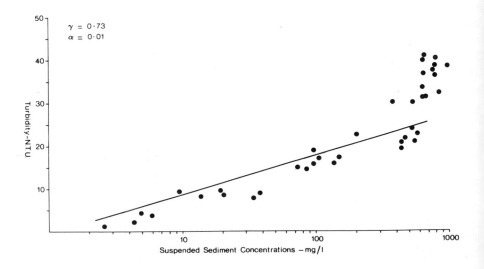

Figure 4. The relation of turbidity and suspended solids

aquatic flora and fauna by the settling of suspended solids, has implications for the stream habitat.

TYPE, SOURCES AND NATURE OF STREET SURFACE PARTICULATES

The bulk components of street surface sediments and their grain size characteristics are shown in Figure 5, the hydraulic implications of the grading curves having been

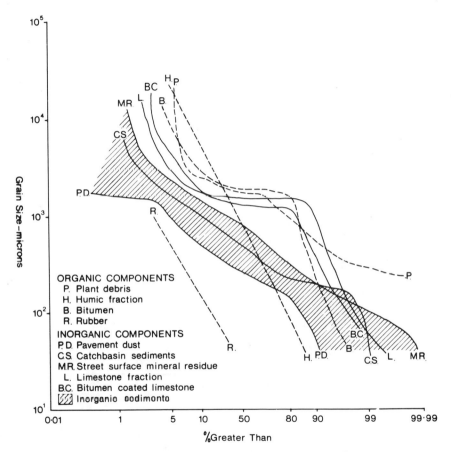

Figure 5. Bulk components of stormwater sediments.

discussed elsewhere (Ellis, 1976). The inorganic mineral fraction, which includes substantial amounts of brick, glass and aggregates, makes up 50-80% of the total sample weight. Organic materials make up a very variable component but can contribute up to a third of the total weight, with the rubber and bitumen fractions typically comprising three quarters of all organics. The association of the organic materials with low particle sizes is seen by inspection of the log normal plots for the humic and rubber fractions (Figure 5). Between 15-70% of these organic components are

found in association with particle sizes of less than
0.06 mm. Volatile solids show a distinct preference for
the lower size ranges (Figure 6). The redistribution of
particle sizes for the above and below ground phases of
stormwater discharge shown in Figure 6 exhibits the
concentration of organics to sizes below 0.06 mm through
physical disintegration by hydraulic action in the sewer
system. The implications for water quality arising from

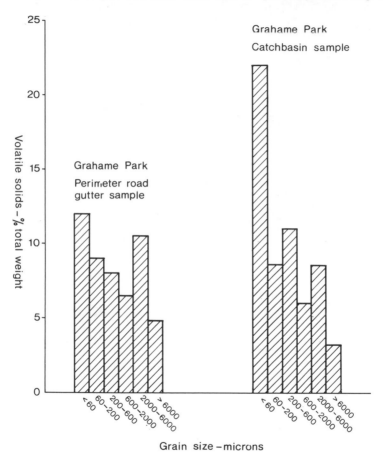

Figure 6. Volatile solids and particle size.

the bulk composition and size characteristics of stormwater sediments stem from the high ionic absorptive capacity
of such accumulated material in receiving streams as thick
benthal sludge. The uptake of cationic species by
decaying organic rich substrates and by clay sized inorganic sediments, as well as their appreciable oxygen
demands, is well documented (Gardiner, 1974; Forster and
Muller, 1973; Ellis, 1976).

Urban stormwater sediments are obviously mixtures
derived from a variety of sources (Figure 7). Roofs and

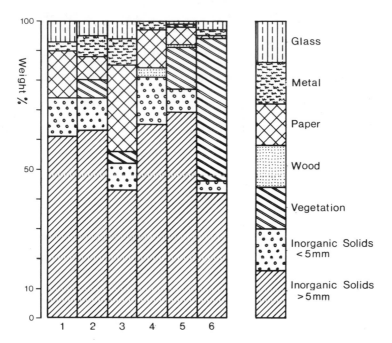

Figure 7. Composition of street surface materials in the Graham Park catchment.

street surfaces provide the bulk of inert solids but also contributo asphalt, bitumon, rubbor and joint filling compounds. Accelerated surface erosion resulting from building and demolition activity, as well as from trampling of grass verges and deflation from vacant plots, will provide a further loading of solids and organics. Consid-erable organic and humic components are derived from leaves, grass clippings and pollen. Traffic produces shreds of tyres, exhaust emissions, oils, plastic, metal and glass, which all accumulate in the gutters. Packaging materials provide a constant input of litter and refuse, whilst food, animal and bird droppings add to the septic qualities of the roadside mix. Spillages of sand and gravel, street car cleaning and winter sanding provide a further source of particulate materials. Figure 7 shows the high proportion of humic material and fine solids emanating from the residential areas which reflects landscaping and gardening work.

Large amounts of organic material contributed to street surface particulates can exert high BOD levels in streams and may damage aquatic life. Lignins and tannins derived from wood fibre and vegetation are very resistant to biological oxidation, whilst grease and oil foul storm drains. Heavy metals can accumulate in the benthal sludges and organisms until lethal limits are reached.

The importance of the fine solids fraction in the total composition of street surface materials is seen from Figure 7 and their pollutional hazards are underlined and specified by consideration of the size distribution of some of the included contaminants (Figure 8). Only 4-8% of the suspended solids on the street surface is composed of material below 0.06 mm in size, yet this fraction accounts

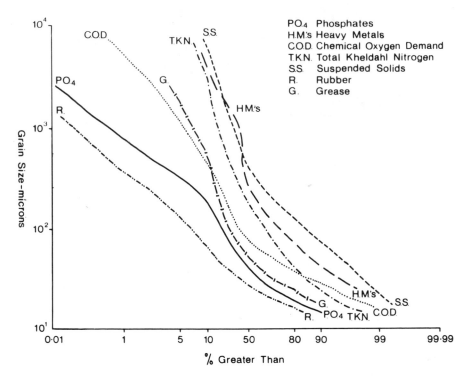

Figure 8. Particle size ranges of street pollutants.

for 25% of the oxygen demand, 30-50% of algal nutrients, 30% of the heavy metals, 50-60% of grease and rubber, and 10% of the total coliforms. Between 40-90% of all pollutants are associated with particles of less than 0.2 mm, with the nutrients and organics being particularly concentrated in the fine fraction. This high concentration must be of considerable importance for eutrophic levels in the receiving stream.

STREET SURFACE PARTICULATES AND CLEANING EFFICIENCY

The distribution of particulate solids across selected
street surfaces within the catchment was determined using
Sartor and Boyd's (1972) method. All sites showed over
80% of the solids within one metre of the kerbside.
Figure 9, for the busy perimeter road of the Graham Park
estate, is typical of most sites. The distribution is
highly skewed, the extent and magnitude of which being
dependent on the incidence of roadside parking. As noted
by Sartor and Boyd (1972), the kerb acts as a protective
barrier against which particulates accumulate on the roadside. Low kerbs, adjacent to unpaved areas which are
either flat or sloping away from the street surface, show
little evidence of solids accumulation and require minimal
routine sweeping. Passing traffic blows the deposited
dust over the low kerb onto the bounding gravel or grass
tracks. During storm events this dust is washed into
sewers but at a lower rate than that from roadside gutters.
    A mechanical rotary sweeper reduces the total solids
loading, although there is some tendency for redistribution
of solids across the roadway which results in some zones
being cleaner prior to the sweeping pass (Table 2).

Table 2.   Concentrations of street surface contaminants.

	BEFORE CLEANING (mg/1)	AFTER CLEANING (mg/1)
Total Solids	6418	4201
Suspended solids	5817	3235
Volatile Solids	3005	1621
COD	3563	1042
BOD	180	60
Total Kheldahl Nitrogen	5.1	1.9

The reductions in loadings are less for the particulates
than for other parameters. Prior to the cleaning run by
the brush sweeper, the arenaceous fractions of between
0.1 - 2 mm are dominant in the grading curves (MRB on
Figure 9). Following a cleaning run there is a marked
redistribution of particle sizes, as shown by a comparison
of both histograms and grading curves. The curves have
breakpoints at about 1 mm and show an increasing removal
efficiency with increasing particle size such that there
is a near total removal of sizes above 10 mm. However,
there is a strong inverse relationship between decreasing
particle size and removal efficiency; the solids fraction
below 0.06 mm being reduced by only 15-20%. Reference to

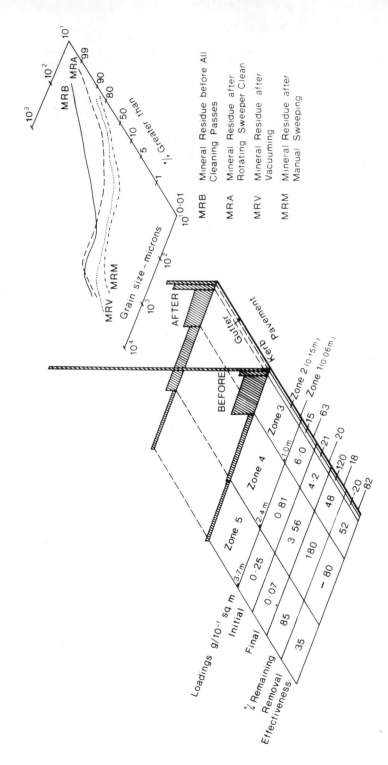

Figure 9. Street surface particulate distribution and cleaning effectiveness.

Figure 8 suggests that the removal of particles greater
than 1 mm, even assuming a 100% cleaning efficiency, would
eliminate no more than 20% of a broad spectrum of contaminants. Clearly, the control of street surface pollutants
in urban areas must depend on a more efficient removal of
a wider range of particle sizes if the shock loading of
urban stormwater discharges on receiving streams is to be
reduced. Manual sweeping, although a less efficient
expenditure of manpower resources from the local authority
viewpoint, does provide a far more effective means of
removal of the finer solids fraction. The grading curve
for the mineral residue left after a manual sweep (MRM on
Figure 9) shows a more effective range of solids removal
except for particle sizes below 0.1 mm. Vacuum sweeping
of the road surface produces slightly better overall
results than manual sweeping (MRV on Figure 9) although
there is very little difference for particle sizes below
0.06 mm. Multiple mechanical sweeping runs do not significantly reduce the yields of fine solids to the sewer
system (Sartor and Boyd, 1972) thus the effect of this
fraction on streams must be buffered elsewhere.

OXYGEN RATIOS AND ORGANIC SOURCES

High ratios of COD to BOD typify urban stormwater (Table 2
and Figure 10) suggesting the inhibition of biological
oxidation by the toxicity of particulate materials. The
COD to BOD ratio for the separate sewer system of Graham
Park ranges from 1.6 to 30, reflecting a significant level
of chemically oxidisable inorganic material. Oxygen
demands increase after wet periods (Figure 10) with COD
increasing faster than BOD. This implies that the organic
oxidisable component of the total solids tends to accumulate on the street surface faster than the inorganic
component. This shows that the sources of the street
surface contaminants are contributing organic matter more
rapidly than inorganic materials. The importance of the
organic loadings contributed from varying land uses
indicated on Figure 7 is thus crucial to the intensity and
pattern of contaminant loadings in urban catchments.
Vehicular inputs such as oils and grease, as well as
organic litter, leaves and vegetation are dominant over
inert, inorganic materials such as sand and dust.
    Figure 10 indicates that fixed sources of materials
containing both organics and inorganics (e.g. the street
surface) are not significant contributors to the total
loading as their curve would plot as a horizontal line.
Whether this conclusion is valid for all types of highway
surface is debatable. Motorway runoff sediments, sampled
from the adjacent Hendon Urban Motorway (Figure 1),
contain between 25-50% mineral residue with 20-40% oils
and 30% rubber and bitumen. The sediments collected by
Hedley and Lockley (1975) from sedimentation chambers on
the Aston Expressway in Birmingham were 65% organics and
also composed largely of oil, rubber, bitumen and grease.
These motorway studies confirm the trends indicated by the
COD/BOD plots of Figure 10, that street surface type,

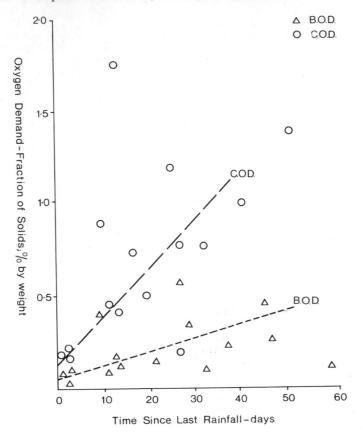

Figure 10.   COD - BOD relations.

texture and characteristics are not of prime importance to pollutional loadings. However, preliminary investigations by Green (1974) on the Ml Motorway near the Watford Gap, quoting pollutant loadings on asphaltic surfaces often 60 to 80% higher than on adjacent concrete surfaces, would place some reservations on the implications of the oxygen ratios until further research can clarify the problem.

CONCLUSIONS AND MANAGEMENT IMPLICATIONS

Pulse loadings of particulate materials from urban catchments to receiving streams during storm events are often of raw sewage quality and the relations of sediment discharge to stormwater flow can be highly variable. The major characteristics of stormwater solids which affect receiving water quality are particle size and composition, with the solids fraction below 0.06 mm and the percentage organic content being of prime importance. Standard street cleaning practice is ineffective in removing the critical

particle size ranges and the significance of this for
eutrophic levels and deoxygenation of the receiving stream
is undeniable. Proper, and more stringent, management of
urban street cleaning procedures are necessary if a marked
reduction in pollution loadings of street surface runoff
is to be effected. Urban sweeping programmes should be
devised to take into consideration such factors as accumulation rates in differing land uses, local precipitation
patterns as well as special short term activities such as
building or demolition. Maintenance of street surfaces
are required to ensure better sweeper collection efficiency
and prevent localised accumulation of particulate
pollutants. Local parking regulations may be necessary to
free access for sweepers, where roadside parking is a
particular problem.

Urban stormwater quality adds another dimension to
flood management for runoff acts as a carrier of pollutants;
public health, recreation and aesthetic considerations
need to be included as well as the more traditional economic
considerations. Thus water pollution management in urban
areas requires that land use and water uses are considered.
To date, urban channels and sewer systems have been
designed to carry stormwater through the catchment to the
receiving stream in the minimum time. With an increased
interest in multi-purpose water systems, which include a
pollution loading, other alternatives of stormwater disposal need to be explored. The most feasible alternatives
appear to be attenuation of peak flows by structural and
land use controls, designed on-line storage and the treatment of flows prior to release into the receiving stream.
McPherson (1972) and Puertner (1976) have described current
systems and techniques implemented in the United States
which combine both flow attenuation by storage and pre-
treatment. Both workers agree on the complexity of the
problem resulting from the instantaneous and frequent
nature of the required on-line control and the combination
of flow relation and treatment capacity. Catchbasins and
detention tanks are not effective in trapping the important
clay-silt fraction or in retaining a significant proportion
of the volatile solids, although heavier solids and
vegetation are trapped. Such detention at outfalls tends
to act as an unregulated digester with the supernatant
being flushed into the receiving stream during the first
stage of stormflow.

Angino et al. (1972) and Ellis (1976) have suggested
possibilities of re-use of stormwater from urban areas, but
such uses require the development of stringent criteria to
distinguish between usable and non-usable runoff as the
flow and pollutant peaks do not necessarily correspond. On-
line storage and treatment can be taken back to the roadside gutter and gully stage of the stormwater system.
Special recessed kerbside gutters might concentrate particulates more efficiently to aid a more efficient removal by
vacuum cleaning or flushing techniques. In series, ultra-
high filtration facilities, perhaps coupled with swirl
separation, might be practised at the gully-pot stage to
separate the fine solids fraction into concentrated side
stream flows for further treatment. Pilot evaluations of

such techniques are required to assess the costs and benefits of incorporating these methods into an overall urban stormwater pollution control plan.

Although urban runoff does not maintain a high loading on the receiving stream for prolonged periods, the impact of the pulse loading on the ecological balance and general habitat of the receiving stream provides a permanent deficit. The biota must not only withstand the hydraulic force of the peak storm discharge but must also recolonise a highly inhospitable environment dominated by thick organic benthal sludge.

ACKNOWLEDGEMENTS

The author is grateful to the Middlesex Polytechnic for providing financial assistance in support of this paper and to the Greater London Council, Public Health Engineering Department and the London Borough of Barnet Surveyor and Engineers Department for aid in the instrumentation of the catchment.

17

THE DISPOSAL OF DOMESTIC AND HAZARDOUS WASTE AND ITS EFFECT ON GROUNDWATER QUALITY

J.D. Mather and A. Parker

Environmental Pollution Section, Institute of Geological Science; and Environmental Safety Group, Harwell Laboratory, Didcot, Oxfordshire.

ABSTRACT

The factors affecting the composition and strength of leachates from landfill sites are evaluated and published work, relating to the formation and migration of leachates, is reviewed. The results of a three year DoE research programme are summarised, emphasis being given to field investigations at existing landfills and to controlled experiments with lysimeters. It is concluded that a correctly managed policy of 'dilute and disperse' is likely to result in less of an overall environmental impact than a policy of 'concentrate and contain'.

INTRODUCTION

The deposition of domestic and hazardous wastes in landfill sites has long been an accepted method of waste disposal. The environmental impact of such sites can be minimised by depositing and compacting wastes in shallow layers which are covered with a layer of earth or inert material at the conclusion of each day's operation. However, if wastes buried in such landfills come into contact with water, either from infiltrating rainfall or from liquid wastes, an obnoxious, mineralised leachate is produced.

The composition and strength of this leachate depend upon three principal factors; waste composition, moisture content and landfill age. The volume of leachate produced depends on the amount of moisture infiltrating the waste and no leachate will migrate from a site until moisture levels exceed the field capacity of the waste.

The composition of the waste buried at a site determines the types of potential pollutant appearing in the leachate. In domestic waste the decomposition of organic matter will contribute soluble organic compounds measurable as BOD, COD and organic nitrogen. Ammonia and carbon dioxide will be released and will contribute to the alkalinity. Metal ions such as those of calcium, magnesium, sodium and iron are common constituents of domestic waste and can be expected in high concentrations in leachate as can the common anions, chloride and sulphate. A typical range in composition of leachates from domestic waste landfills is given in Table 1 (L.S. Wegman Co. Inc. quoted by Geswein, 1975).

Table 1: Typical domestic refuse leachate composition

Component	Range of Values*		
pH	3.7	-	8.5
Conductivity	100	-	1200
Biochemical Oxygen Demand (BOD)	7050	-	32 400
Chemical Oxygen Demand (COD)	800	-	50 700
Alkalinity	310	-	9500
Calcium	240	-	2570
Magnesium	64	-	410
Sodium	85	-	3800
Potassium	28	-	1860
Total Iron	6	-	1640
Chloride	96	-	2350
Sulphate	40	-	1220
Phosphate	1.5	-	130
Organic Nitrogen	2.4	-	550
Ammoniacal Nitrogen	0.2	-	845
Suspended Solids	13	-	26 500

* Values in milligrammes per litre, except for pH (pH units) and conductivity (micro mhos per cm.)

The prediction of leachate components from industrial waste requires a detailed knowledge of the wastes' composition. It might be anticipated that much of what enters the landfill as waste would eventually appear in the leachate. However, this will depend very much on such factors as solubility, oxidation/reduction potential and pH; more field and laboratory data are required before confident predictions can be made.

The strength of leachate increases with increased moisture content of the waste. The moisture may be contained within the waste when it is emplaced or may subsequently infiltrate the waste through precipitation. A probable pattern of decomposition of domestic refuse in response to changes in moisture content has been identified (Farquhar and Rovers, 1975). At low moisture content local aerobic conditions are possible yielding low concentrations of organic material in the leachate, a neutral to slightly alkaline pH and conditions favourable for metal precipitation. As moisture content increases towards saturation anaerobic conditions develop increasing the soluble organic content of the leachate. The pH and oxidation/reduction potential are likely to be reduced, increasing the chances of solubilization of metals.

In both experimental and field situations, leachate strength has been found to reduce with the age of the refuse (Anon., 1961; Farquhar and Rovers, 1975). This is because the more easily released materials are discharged quickly so that the quantity of soluble material in the refuse becomes depleted with time.

Leachate generated within a landfill will eventually migrate away from the site at a rate which depends primarily on the hydrogeological conditions. This migration can result in the pollution of both surface and groundwater unless leachates are diluted and degraded by natural processes, or are contained within the site, collected and treated. There is now an extensive literature on this topic which, in the United Kingdom, extends from the early work of Woodward (1906) to the 'Landfill Research Project' currently being undertaken for the Department of the Environment by staff of the Institute of Geological Sciences, the Harwell Laboratory and the Water Research Centre (Anon., 1975). This paper summarises this work and discusses the management implications of the results.

REVIEW

The problem of groundwater pollution from landfill sites has received a considerable amount of international attention over the past 20 years. Investigations around existing sites, in the United States (Anon., 1954b; Anderson and Dornbush, 1967; Fungaroli and Steiner, 1971; Hughes, Landon and Farvolden, 1971; Apgar and Langmuir, 1971), Germany(Golwer and Matthess, 1968; Matthess, 1972), France (Affholder et al., 1973) and other industrialised countries, have indicated the type and extent of pollution to be expected from the deposit of domestic waste. However, there are few detailed published investigations around existing industrial waste landfills (Lieber et al., 1964; Hopkins and Popalisky, 1970) and the behaviour of many potentially polluting species is poorly understood.

Investigations around existing landfill sites have been supported by detailed laboratory studies into the migration potential of various leachate components. Such studies have generally involved the use of columns of remoulded soils and unconsolidated formations which are irrigated with real or synthetic leachates (e.g. Fuller and Korte, 1976; Houle et al., 1976; Griffin et al., 1976 and 1977). Of particular interest is the work of Farquhar and Rovers (1976) in which soil samples were exposed to leachates in flasks and shaken for increasing periods of time. They were able to show that these dispersed soil experiments could be used to give an estimate of the attenuation capacity of a particular formation in respect of a particular leachate.

In the United Kingdom detailed research into ground-water pollution from solid wastes seems to originate with the work of Jones and Owen in Manchester in 1931-4 (reported in Bevan, 1967). The object of this research was to obtain scientific data on the behaviour of the contents of a land-fill after tipping and sealing. This work was followed by the report of the Ministry of Housing and Local Government Technical Committee on the Experimental Disposal of House Refuse in Wet and Dry Pits (Anon., 1961). This report gave details of experiments designed to give information about the risk of polluting groundwater by tipping house-hold waste directly into standing water in hydraulic continuity with groundwater and where leachate derived from tips had access to groundwater. Both small and large scale experiments were performed as well as some on the purification of contaminated water by filtration through sand and gravel. The report also gave examples from the literature of pollution problems associated with landfill sites, and confirmed that untreated household waste should not be deposited in fissured sites from which leachates could escape into groundwater with only minimum attenuation. It also showed that sand and gravel acted as an effective filter for bacterial and organic contaminants but that attenuation of chloride and sulphate was minimal.

In 1969 a research panel of the Society of Water Treatment and Examination and the Institution of Water Engineers (Waterton et al., 1969) published the results of a survey of possible groundwater pollution from landfills situated on various aquifers e.g. Lower Greensand, Triassic Sandstone, Magnesian Limestone and Chalk. Their findings suggested that there was little or no detectable evidence of pollution of groundwater from decomposition of household waste. Following this report Gray, Mather and Harrison (1974) presented the preliminary results of a multi-authority review of the water pollution hazard rep-resented by 2494 landfill sites in England and Wales. Of these, desk studies suggested that only 51 represented a serious pollution risk to major or minor aquifers. It was noted that despite the large number of sites the number of incidents in which groundwater supplies have been significantly polluted remains negligible.

Over the last few years the major interest in the United Kingdom has concerned the disposal of toxic and hazardous wastes to landfill and reports of three inves-tigations have been published (Billington and Tester, 1973; Aspinwall, 1974; Gray and Henton, 1975). In 1964 a Technical Committee was set up to report on the disposal of

solid toxic wastes. Its report (Anon., 1970) gives
examples of pollution of both surface and groundwater by
hazardous chemicals in the U.K. No significant pollution
of underground potable water was reported although some
minor aquifers had been affected.

CURRENT RESEARCH IN THE UNITED KINGDOM

The Technical Committee on the Disposal of Solid Toxic
Waste noted that insufficient scientific research had been
carried out into the disposal of toxic waste (Anon., 1970).
Soon afterwards accounts appeared in the press of alleged
cyanide dumping and these were followed by the introduction
of the Deposition of Poisonous Waste Act, 1972. The
Department of the Environment decided that further basic
information was required on the behaviour of hazardous
wastes in landfill sites before detailed guidelines could
be issued to the industry and in 1973 a three year
research programme was initiated which involved staff from
the Water Research Centre, the Institute of Geological
Sciences and the Harwell Laboratory. The joint programme
covered field investigations at existing landfill sites,
supplemented by laboratory and pilot-scale studies and
controlled field experiments into the movement of
pollutants through the unsaturated zones of various
geological strata to provide information on removal,
attenuation and delay mechanisms.

Field investigations at existing landfills

Twenty sites have been investigated in detail. These were
selected so as to cover the main United Kingdom geological
formations with particular emphasis on major aquifers e.g.
Bunter Sandstone, Carboniferous Limestone, Chalk and Lower
Greensand.
   A wide range of industrial wastes, some containing
potentially toxic materials, was studied including heavy
metals, cyanide, acids, phenols, PCBs, solvents and mineral
oils. Inevitably, each site investigated contained several
types of waste (e.g. domestic, industrial and hazardous)
but information on behaviour of toxic materials obtained at
one site could normally be applied to others.
   After a preliminary site survey, a number of boreholes
were drilled both through the landfill and around the
periphery of the site. Initial boreholes were sited in
the light of existing hydrogeological knowledge but once
the direction of the hydraulic gradient had been established the majority of the boreholes were drilled down-
gradient of the site to enable the extent of any pollution
plume to be established. Analysis of core material and
groundwater enabled the extent of attenuation of any liquid
escaping from the landfill to be determined. Occasionally
boreholes were also drilled through the landfill into
underlying strata so that rock cores could be extracted to
measure the extent of downward penetration of landfill
leachate. Once the main drilling programme was complete,

groundwater monitoring was continued to determine any long term deterioration in quality. The following examples serve to illustrate the type of study which has been carried out and the results obtained.

Wastes containing heavy metals are commonly incorporated into landfill sites. Examples of such wastes include metallic sludges from electroplating processes, paint wastes, lamps, batteries and contaminated demolition waste. In many of these materials the metals are present as insoluble compounds and in general the near neutral pH of a landfill ensures that the heavy metals do not readily enter solution. However, if significant quantities of acids are added, then the pH falls and landfill fluids can contain high concentrations of metal cations. Such a site has been investigated in the Midland Valley of Scotland. This landfill, in a disused quarry within the Carboniferous Lower Productive Coal Measures, is situated above old coal workings which allow the rapid transmission and storage of large volumes of groundwater. A wide range of industrial waste has been deposited since 1962 with a gradual increase in the percentage of liquid waste until 1975. A liquid lagoon is present which consists of an oil/tar sludge overlying an aqueous phase. Owing to the discharge of large volumes of liquid waste containing sulphuric acid the pH of the lagoon aqueous phase is 1-2 with significant concentrations of iron (1000 mg/l), zinc (50 mg/l), chromium (30 mg/l), nickel (15 mg/l), manganese (15 mg/l) and copper (5 mg/l) together with TOC (1300 mg/l) and phenol (20 mg/l). Interstitial fluids from the sandstone bedrock immediately below the sites showed a similar composition but with greatly increased concentrations of certain species. Liquids from the lagoon percolate into the solid waste so that the lower 6-8 m is saturated. It then moves laterally within the sandstone fissures or vertically via fissures or old mine shafts into the workings beneath. Lateral movement of pollutants, although with considerable dilution, was found some 80 m from the landfill.

The detailed analytical work undertaken suggests that the metal cations and the acid solutions could be attenuated by a process involving acid-base reactions, precipitation, dilution and oxidation as shown opposite.

In comparison with the aqueous liquid wastes, the oily wastes are not significantly attenuated in their movement from the quarry within the old workings. The oil remains as a separate phase, floating on top of the underlying groundwater.

Phosphate sludges containing zinc and nickel were one of the principal wastes deposited between 1948 and 1974 at a disused gravel pit overlying Lower Chalk on the western margin of the Chilterns. A detailed investigation showed that the metal sludges remained within the landfill environment and there was no evidence of solubilization of zinc or nickel. The domestic and solid industrial wastes which were also deposited at the site yielded leachates with a considerable pollution potential. However attenuation of these leachates within the unsaturated Chalk beneath the landfill was extremely effective and, at the water table,

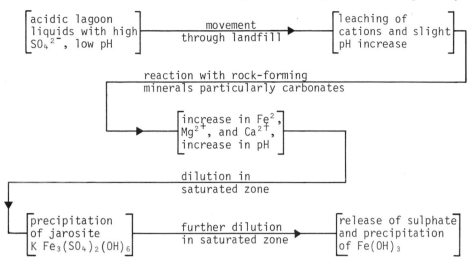

water quality deterioration was confined to an increase in
total hardness producing a 'hardness halo' around the
landfill. Attenuation of the organic component of the
leachate was complete at the Chalk water table and there
was no comparable 'organic halo'. At this site there was
little evidence that fissure systems have enabled significant
volumes of leachate to by-pass the unsaturated zone and
move rapidly to the water table. Thus locally the Lower
Chalk allows slow leachate migration and significant
attenuation (Mather, 1976).
    Phenols are undesirable in groundwater because of the
taste which is imparted by even trace quantities on
chlorination. Industrial wastes often contain phenol
which may appear in landfill leachate. Recent research
work has shown that attenuation of phenols by adsorption
and biodegradation within landfills is slow and so considerable care must be taken in the disposal of phenolic
wastes. Water soluble phenolic contaminents have migrated
into groundwater at a landfill in the Midlands situated in
a series of disused sandpits excavated in glacial sands
overlying thick glacial clays which in turn overlie Keuper
Marl. Since 1971 large volumes of liquids containing
cutting oils and phenols have been deposited which have
migrated within the sands. The sands are primarily a
dispersive medium and require a high groundwater flow rate
to effectively dilute the wastes from the lagoon. However,
the regional hydraulic gradient is low and the head of
liquid in the lagoon is sufficient to produce a
'malenclave' of polluted groundwater down the hydraulic
gradient from the landfill. This is an example of a poorly
sited landfill where adverse hydrological factors result in
the formation of a significant pollution plume.

Controlled field experiments and laboratory investigations

For the optimal siting of landfills it is necessary to know
the extent of attenuation of dissolved pollutants in
various leachates by the underlying strata before they
reach potential groundwater resources. Of particular
interest is the unsaturated zone immediately beneath a
landfill as it is here that the first contact occurs
between leachates and rock minerals. Because of the diversity of groundwater flow conditions, pollutant movement
within three different lithological types - chalk, sand and
gravel - has been studied. The techniques used include
laboratory columns of disturbed materials and field
lysimeters.
  For the investigations in unsaturated sand, representative of intergranular groundwater flow conditions,
lysimeter techniques are being used. These lysimeters
include both monolithic cores containing approximately
0.64 m^3 of sand and field lysimeters which isolate
approximately 50 m^3 of sand. The object of using field
lysimeters is to isolate representative bulk samples of
sand in an undisturbed condition so that pollutants will
move under as near as possible field conditions. The
effects of permeability variations between sand layers,
some rich in clay minerals, and any discontinuities can
then be assessed and a realistic picture built up of
unsaturated zone flow.
  The geological requirements for a field lysimeter site
are an approximately 3 m thick layer of sand underlain by
an impermeable clay. A series of sand masses can be
isolated by the construction of impermeable vertical walls
through the sand into this underlying clay. Clearly the
walls need to be constructed so that disturbance of the
sand is kept to a minimum and the continuity between sand
and walls must be such that edge effects are insignificant.
  A site has been located on an outlier of the Lower
Greensand unconformably overlying Kimmeridge Clay. Four
lysimeters have been constructed, two on either side of
a central trench which allows access to one wall of each
of the lysimeters. Liquid samples from the unsaturated
sand are obtained using suction probes installed almost
horizontally at various depths beneath the surface (Black
et al., 1976).
Three lysimeters are being irrigated with artificial
leachates. These consist of:
(i)   a solution containing heavy metal ions, alkali
metals, alkaline earths, chloride, lower straight chain
carboxylic acids and phenol;
(ii)   phenol, 2:4 dichlorophenol and aniline at a nominal
100 mg/l concentration together with acetate;
(iii)   a solution containing various anions e.g. sulphate,
chloride, acetate, nitrate, phosphate and bromide.
  As an example of the type of information obtained,
concentrations of species in the uppermost suction probe
installed in the lysimeter irrigated with heavy metals in
carboxylic acids are shown in Figure 1. This indicates
that approximately 500 days were required before the most
mobile of the heavy metals studied, i.e. nickel, reached

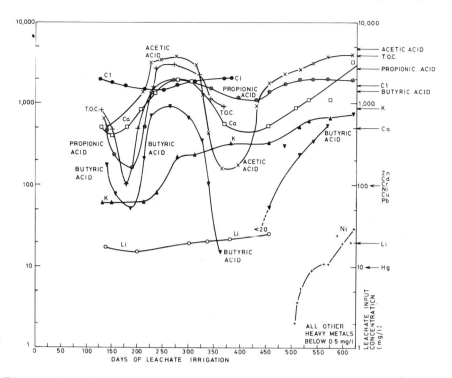

Figure 1.   Concentration of chemical species (mg/l) 400 mm
   below the surface of a sand filled lysimeter being
   irrigated with leachate (see text for details).

a depth of 400 mm below the surface while there was no
evidence of the breakthrough of any other heavy metals.
Even after 600 days, the nickel concentration had only
risen to 30 mg/l compared with 100 mg/l in the input.
These results demonstrate that heavy metals are effectively
retained in the upper layers of the lysimeter, behaviour
which contrasts markedly with chloride which is attenuated
only by dilution and dispersion and attained its input
concentration 100 days after irrigation commenced.   Measurements have shown that the downward velocity of the liquid
front through the lysimeter is approximately 5 mm per day
which is close to the velocity of the chloride front.   The
behaviour of potassium is interesting since even after 600
days irrigation it has not attained its input concentration.
This unexpected result was confirmed by measurements at
the suction probes at lower levels and it is probable that
attenuation is due to incorporation into the biomass
formed by microbial reactions within the lysimeter.   Both
acetic and butyric acid concentrations show marked
decreases after about 400 days irrigation followed by
increases toward input concentrations.   The decrease is
associated with increased temperatures and microbial
activity during summer months resulting in more efficient
decomposition of these carboxylic acids.   The effect is not

so marked with propionic acid which is more resistant to biological attack.

MANAGEMENT IMPLICATIONS

There are two major options available in the selection of landfill sites. The first is to select sites which allow leachates to migrate away at slow rates so that natural processes can attenuate and dilute them before they reach potential or developed water resources. This is the policy of 'dilute and disperse' and is normally accompanied by measures to limit the volume of liquid entering a landfill. The second strategy is to contain leachates within the landfill either by siting the landfill on a poorly permeable formation such as a clay or shale or by providing an artificial impermeable liner such as bentonite or a synthetic polymeric membrane (Mather, 1977). This is the policy of 'concentrate and contain' and needs to be backed by a system of leachate collection and treatment.

The main objection often put forward to dilute and disperse is that insufficient data are available to assess the attenuation capacities of potential landfill sites. However the international work discussed in this contribution has resulted in the accumulation of a considerable volume of data on the various attenuation processes which affect leachates, particularly those arising from domestic wastes. The fact that very few documented cases of significant groundwater pollution caused by landfills in the U.K. have been reported, despite the long experience of the use of this technique for waste disposal, is significant and supports the view that natural processes are available beneath landfills and are effective in attenuating leachate concentrations. As a result of the work carried out in recent years, particularly in the DoE programme, a much better basis for understanding the mechanisms involved has now been established.

By correctly siting landfills it is possible to provide conditions favourable to the development of attenuation processes. The data available also permit preparation of general guidelines for the disposal of domestic wastes so that problems of ground and surface water pollution are avoided (Mather, 1976, 1977). Over most of the United Kingdom, landfill sites operating on the dilute and disperse principle can be identified which offer excellent prospects for effective domestic waste leachate attenuation so long as basic hydrogeological principles are followed in site selection. The effectiveness of natural processes can often be enhanced by using material such as crushed limestone or concrete demolition wastes to line sites. As well as domestic wastes many industrial wastes can also be disposed of at dilute and disperse sites and such a technique is particularly beneficial for those wastes whose leachates are comparable with those from domestic wastes.

Sites which rely on containment of leachates have a number of disadvantages compared to dilute and disperse sites. Leachates must be collected and removed otherwise sites become waterlogged resulting in gas problems and the eventual pollution of surface water. The collected leachates have to be disposed of or treated in some way and this can be expensive. Covering waste with an impermeable blanket to prevent the formation of leachates results in deceleration of decomposition and biodegredation so that wastes are effectively stored for long periods. This results in problems in the reclamation and reuse of such sites once tipping has ceased.

Current research in the United Kingdom suggests that although some hazardous wastes may require containment many others can be safely dispersed to the environment. Large volumes of industrial liquid wastes may result in extensive lagoons of contaminated liquid if they are discharged at containment sites. Such wastes can be discharged to specially selected dilute and disperse sites within strata containing limited groundwater resources where some pollution would not result in environmental problems. Taking a long term view, it would seem that a correctly managed policy of dilute and disperse is likely to result in less of an overall impact to the environment than a policy of concentrate and contain.

ACKNOWLEDGEMENTS

Part of the work on which this paper is based has been funded by the Department of the Environment, although opinions expressed are those of the authors alone. The paper is published by permission of the Director of the Institute of Geological Sciences and of the Harwell Laboratory, United Kingdom Atomic Energy Authority.

18

TEMPERATURE CHARACTERISTICS OF
BRITISH RIVERS AND THE EFFECTS OF
THERMAL POLLUTION

K. Smith

Department of Geography, University of Strathclyde

ABSTRACT

*The effects of human activity on river water temperatures
are examined. It is shown how unmodified thermal
behaviour can be related to local hydrometeorological
conditions and a basic distinction is made between the
temperature characteristics of streams and rivers. Some
emphasis is placed on the causes and effects of thermal
modification which, in different situations, can reach
$12^{o}C$. Employing temperature range as the most sensitive
index it is concluded that heated effluent discharges and
probably urbanization tend to increase the range of
thermal variation in lowland rivers, whilst the reverse
effect occurs in upland rivers due to reservoir releases
and afforestation.*

INTRODUCTION

The thermal regime of a river is one of the most important
of all water quality parameters and has both an ecological
and an economic significance. According to most freshwater biologists, such as Reid (1961) and Macan (1974),
water temperature is the primary environmental factor
influencing the distribution and well-being of aquatic
communities. At the same time, the thermal characteristics
of the larger rivers have an important economic role as a
water source for industrial cooling purposes. In Britain
the main demand comes from the electricity generating
industry where, according to Langford (1972), even the most
efficient power stations have to dispose of 50-65 per cent
of all generated energy as heat in cooling water. Over 50
per cent of Britain's electricity is now produced at power
stations sited on rivers and the Central Electricity
Generating Board takes for cooling approximately half of
all the licensed water abstractions in England and Wales,

almost all of it from surface sources (Water Resources Board, 1973). River water temperature can affect the costs of thermal power generation, because the efficiency of the steam cycle diminishes with a rise in condenser temperature, and Clark and England (1963) calculated that a difference in inlet temperature from 15°C to 20°C would increase the annual coal consumption of a 2000 MW base load station by 100 000 tonnes.

The ecological and economic aspects of water temperatures become inseparable as a result of the thermal modifications caused by the discharge of heated effluents. Such thermal enrichment is the most important man-made impact in Britain since the demand for cooling water, especially from the power industry, is large in relation to the capacity of our rivers. Thus, Hawes (1970) has stated that generating stations require almost 3.5 m^3/s of cooling water per 100 MW generated. Similarly, Ross (1959) claimed that a comparatively small 400 MW unit has to dispose of some 2000 million BTU/hr. This would be sufficient to raise the temperature of a flow of water of 19 m^3/s, which is approximately double the dry weather flow of the Thames at Teddington, through 5°C. The overall demand is very large. Indeed, the total amount of fresh water circulated in power stations probably exceeds 40 per cent of the mean annual runoff and, during low flows, there is insufficient water in any British river to cool even one 500 MW unit by direct abstraction (Hawes, 1971). Since new power stations exceed 2000 MW capacity, with a gross cooling requirement around 65 m^3/s, it is clear that such demands can only be met by indirect means and the recirculation of water through cooling-towers. Using this method such large stations need abstract only 3 per cent of their total requirements from rivers in order to replace the 1 per cent evaporative loss and the 2 per cent returned to the river to allow for make-up and the prevention of an over-concentration of solids (Ross, 1970). Alabaster (1969) indicated that about half the stations in operation used the direct system, nearly one-third had full cooling-tower capability with the remainder using a combined system depending on river flow.

Although the effects of condenser water released by the power industry have attracted most attention, the human impact on river temperatures is by no means confined to such influences. There is growing evidence to suggest that river temperatures are sensitive to a variety of inadvertent modifications, some of which relate to the lowering of temperatures and to man-made effects in essentially rural environments. There are, however, two fundamental difficulties surrounding the interpretation of all such modifications.

The first problem is that our understanding of thermal modification is greatly impaired by an incomplete knowledge of temperature conditions in natural rivers. At a reasonably conservative estimate there are now probably over 2000 publications dealing with the relationships between water temperature and fish and other aquatic organisms. This is somewhere between 10 and 100 times greater than the body of literature on the physical mechanism controlling natural water temperature variations.

The second difficulty is that, despite this wealth of literature, many of the specific effects of river temperature on fresh-water ecosystems remain unsolved. There are many reasons for this. For example, although many studies have attempted to isolate the ecological role of temperature, thermal conditions are so inter-related with other environmental variables that temperature can rarely be seen as a single limiting factor. Another reason is undoubtedly because most fresh-water ecologists have concentrated on the effects of temperature on fish populations and have thereby neglected its influence on associated organisms in the food chain (Allan, 1969).

The significance of these deficiencies in understanding is that they make it difficult to identify thermal modification and almost impossible to provide a wholly acceptable definition of thermal pollution. It would be unrealistically severe to equate pollution with modification, not least because many existing thermal alterations have not resulted in clear-cut effects on aquatic life. On the other hand, it is well established that the main ecological consequence will be the impoverishment of natural biological communities and, as a working definition of thermal pollution, Cairns (1967) has proposed that it should denote an environmental change which reduces the species diversity for a particular locality by more than 20 per cent from the empirically determined level. Of course, even such radically changed environments could be deemed to have certain advantages. For example, Iles (1963) has claimed that heated discharges could be used beneficially in domestic heating and in the rearing and farming of both sporting and commercial fish in Britain. Therefore, in view of these problems, the term modification is preferred to the use of the word pollution in this contribution.

NATURAL RIVER TEMPERATURES

The systematic recording of river water temperatures in Britain dates back to measurements made in 1937 at Northampton and Oundle on the rive Nene (Herschy, 1965). Traditionally, standard observations have comprised daily maxima and minima, and averaged monthly data from representative stations have appeared in the *Surface Water Year Book of Great Britain* since the publication of the 1953-54 volume. The *Year Book* for 1971-73 lists a total of 68 stations. Despite recent increases, the density of temperature measuring stations still compares unfavourably with the distribution of other hydrological measurements and, with the growing tendency to incorporate temperature measurement into general water quality monitors, by no means all sites now record daily maximum and minimum values.

These standard hydrometric data have attracted little analytical attention from hydrologists. As shown by Smith (1972), most investigations have been undertaken by biologists who, in order to make their studies more

ecologically meaningful, have concentrated on empirical,
short-term studies of small, upland streams. In comparison,
there have been few physically-based studies relating water
temperatures to hydrometeorological conditions. On
theoretical grounds, river temperature variations are best
determined from heat transfer processes, which can be
approached through energy budget analysis. However, in the
absence of detailed energy balance data, attempts have been
made to represent the net changes of heat storage in rivers
through the more accessible parameter of air temperature
(Smith, 1968). The effectiveness of this method has been
demonstrated in north-eastern England by Smith (1975) and
Smith and Lavis (1975). As shown in Table 1, the highest
correlations were obtained for the headwaters rather than
the downstream reaches and for maximum rather than minimum
values, although all results were statistically
significant at the 0.1 per cent level. Overall, it was
found that air temperatures accounted for between 59 and
92 per cent of the variation of daily maximum and minimum
river temperatures.

Table 1. Correlation between daily air and water
          temperatures for selected stations on
          the rivers Wear and Tees.

River	Station	Approx. Mean Flow	Temperature	r	r^2
Tees	Broken Scar	$18.0 m^3/s$	Maximum	0.82	0.68
			Minimum	0.78	0.60
Tees	Moor House	$0.5 m^3/s$	Maximum	0.92	0.84
			Minimum	0.77	0.59
Wear	Lanehead	$0.03 m^3/s$	Maximum	0.96	0.92
			Minimum	0.91	0.82

On a national scale, river water temperatures show a
decrease in annual mean values with increasing latitude and
altitude. There is also a well-defined seasonal cycle,
with temperatures rarely exceeding $25^{\circ}C$ or falling below
$0^{\circ}C$. These features are illustrated in Figure 1A, which
compares the mean monthly temperatures during 1964-65 for
the Tees at Dent Bank (227 m OD) with those for the Great
Stour at Chartham Old Mill (c. 12 m OD). The greater
seasonal amplitude of the Tees is highlighted in Figure 1B
by means of a dimensionless coefficient which expresses
mean monthly temperature as a ratio of the annual value.
These coefficients, ranging from 0.18 to 1.92 for the Tees
compared to only 0.64 to 1.39 for the Great Stour, suggest
that thermal behaviour is partly dependent on local catch-

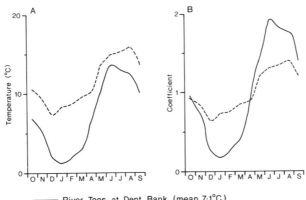

———— River Tees at Dent Bank (mean 7·1°C)
----- River Great Stour at Chartham Old Mill (mean 11·4°C)

Figure 1A. Mean monthly temperatures (1964-65) for the
Tees and Great Stour.
1B. Coefficient of mean monthly temperature
1964-65.

ment characteristics. In this case it would seem that the
comparatively large ground-water contribution to the Great
Stour exerts a stabilising effect on temperatures.
Some typical relationships between air and river temperatures are depicted in Figure 2, which relates to the
lower river Wear in north-east England. At Sunderland

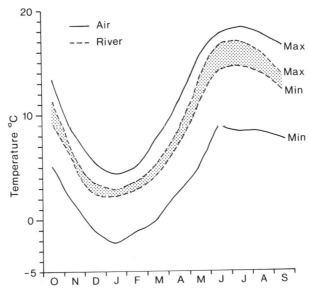

Figure 2. Mean monthly temperature range (1962-66) for the
Wear at Sunderland Bridge compared with air temperatures
at Houghall (after Smith, 1968).

Bridge the Wear has a mean flow of approximately 10 m³/s
and it can be seen that average river temperatures are
consistently higher than air temperatures recorded 4 km
downstream. The range of river temperatures is much lower
than the equivalent air temperatures owing to the thermal
capacity of the water and the fact that, unlike the atmosphere, river temperature virtually never falls below 0°C.
However, it is also apparent that during the summer both
the range of river temperature variation and the maximum
values increase markedly in comparison to the air
temperatures.

Interesting relationships due to differences in flow
volume emerge when air and water temperatures are examined
for the downstream profile of a major river, as reported
for the Tees by Smith (1975). For example, Figure 3
compares the average diurnal temperature variation at the
headwater station of Moor House (533 m OD; average flow
0.5 m³/s) with that at the lowest gauging site of Broken
Scar (37 m OD; average flow 18 m³/s) during a week of
summer anticyclonic weather. Firstly, it can be seen that,
although the mean air temperature range is slightly
greater at Moor House than Broken Scar, the river temperature range was almost three times as high at 14°C
compared with only 5°C at the lower station. Secondly,
Moor House also recorded the highest daily river values
and it is commonly found that, despite lower air temperatures upstream, the highest absolute maxima are observed
in the upper basin. Thirdly, Figure 3 illustrates the

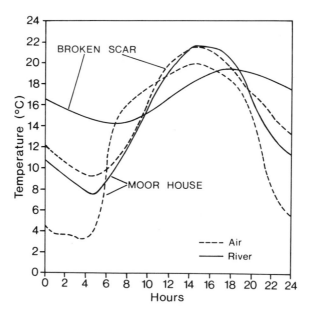

Figure 3. Mean hourly variation of river and air temperatures on the Tees at Moor House and Broken Scar
between 7-13 June 1969 (after Smith, 1975).

well-established increase in phase-lag between air and
river temperatures with distance downstream. Thus, whilst
the diurnal air and water temperature cycles are almost
coincident at Moor House, a lag of 2-3 hours occurred on
the lower reaches. All these features may be attributed
to the decrease in thermal capacity of rivers associated
with smaller upstream dishcarges.

In the absence of any national classification of river
temperatures, a basic distinction can at present only be
drawn between the thermal characteristics of streams and
rivers. This distinction can be sustained since the
smaller flows of streams make them more responsive to the
heat exchange processes affecting any water surface. In
practice, this produces a more variable temperature
regime which is sensitive to both prevailing weather conditions and to differences in the local topographic
stream environment which, in turn, often influences downstream thermal changes. Most of the information on streams
relates to small upland catchments often less than 1 km^2
with flashy flows averaging less than 50 l/s and dry
weather flows falling below 3 l/s. These streams have a
low heat storage capacity, with rocky beds exposed to the
direct radiative fluxes, and therefore show large absolute
temperature ranges. Minima frequently fall to 0°C because
of the altitude whilst annual maxima normally attain 25°C,
and the highest natural maximum of 28°C was recorded by
Macan (1960) in a small Lake District stream. By comparison, as indicated in Table 2, a much larger river
would normally have an annual maximum around 20°C.

Table 2. Typical thermal characteristics of streams and
rivers.

Characteristic	Streams	Rivers
Annual maximum	25°C	20°C
Annual minimum	0°C	1°C
Diurnal range (winter)	2°C	<1°C
Diurnal range (summer)	10-12°C	3°C
Maximum rate of change	3°C/hr	0.5-1.0°C/hr

Thus, the Severn at Ironbridge, with low flows approximately 6 m^3/s, produced an absolute maximum of only 22.8°C
over a decade of measurement (Langford, 1970). Similarly,
the Severn exhibited virtually no diurnal variation in
winter and a maximum daily range of only 2.5°C in summer.
On the North Tyne, a rather smaller river, Boon and Shires
(1976) have found diurnal variations averaging over 1°C in
winter and 3.3°C in summer, with an absolute daily range
of 7°C. However, the smallest streams can occasionally

produce diurnal fluctuations up to $4^{\circ}C$ or $5^{\circ}C$ in winter and up to about $15^{\circ}C$ in summer.

Such small streams are also clearly influenced by the micro-environment of the drainage basin. Thus a series of independent experiments in the uplands of northern England by Macan (1958), Edington (1966) and Crisp and Le Cren (1970) have respectively demonstrated the effect of steep banks, bank-side vegetation and a subterranean stream section through limestone in shading the stream and thereby depressing water temperatures. These effects are most apparent during anticyclonic weather in summer when solar radiation receipts are high and streamflow is low. For example, Smith and Lavis (1975) have noted that ground water seepage from a small aquifer can reduce stream temperatures by $4-5^{\circ}C$ over a distance of 300 m under these conditions.

TEMPERATURES IN MODIFIED RIVERS

Essentially, the thermal modification of rivers occurs as a result of either discharges, with temperatures higher or lower than ambient, being introduced to watercourses, or land-use changes which alter the local heat transfer processes sufficiently to influence the river's energy budget. In practice, heated effluents are the major problem for lowland rivers in urban areas, whilst upland reaches are increasingly subject to the thermal impact of reservoir releases. Land-use effects have so far been monitored successfully only for the smaller streams in rural environments but it is likely that many modifications have occurred, especially owing to direct changes in the radiative fluxes.

Heated Effluents

The main source of heated effluents is direct-cooled power stations but most urban areas produce a wide variety of similar releases. For example, in a survey of the lower Thames, Gameson et al. (1957) showed that heat rejected from thermal power stations was responsible for 74.5 per cent of the artificial heat increment as opposed to 9 per cent from sewage effluent, 6.5 per cent from freshwater discharges, 6 per cent from industrial discharges and 4 per cent from biochemical activity. The water passing through the condensers is normally warmed by $8-9^{\circ}C$, but the effects depend on many factors including the dilution ratio, the degree of mixing, the temperature and velocity of the river and the operating characteristics of the power station. Thus, in general terms, it is important that in Britain the major electricity demand is in the winter when high river flows and low temperatures prevail.

Some of these features can be illustrated by Ironbridge 'A' power station, which has a maximum output of 210 MW and normally uses between 5 and 50 per cent of the mean daily flow in the Severn. As a result, downstream temperatures have been raised between $0.5^{\circ}C$ during spates to $8^{\circ}C$ during

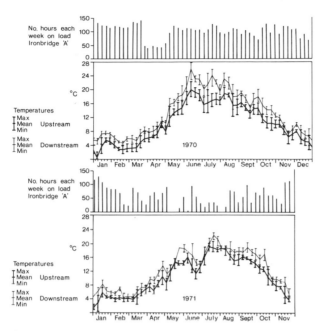

Figure 4. Weekly maximum and minimum temperatures on
  the Severn upstream and downstream of Ironbridge
  power stations in 1970 and 1971 in relation to
  operating hours at Ironbridge 'A'. (After Langford
  and Daffern, 1975).

low flows (Langford, 1970). The longer-term effects are
shown in Figure 4 which plots weekly upstream and downstream temperatures during 1970 and 1971 in relation to
station operating hours. It can be seen that, due to the
decreased operation in 1971, the effluent had a smaller
overall effect on river temperatures than in the previous
year, although individual daily increments did occasionally
attain 7°C (Langford and Daffern, 1975).   The cumulative
impact appears greatest in winter, and Langford (1971) has
claimed that the downstream reaches may be 3-5 weeks in
advance of upstream in accumulating degree-hours (above
0°C) at that season compared to up to 3 weeks during
summer.
     Several British rivers are now subject to large-scale
thermal modification by heated effluents. For example,
Lester (1967) has reported that the mean annual maximum
temperature of the middle and lower reaches of the Trent
had been raised by about 4°C. Individual power stations
on this river appear to increase temperatures immediately
downstream by about 10°C according to data presented for
Drakelow by Langford and Aston (1972) and for Castle
Donington by Aston and Brown (1975). On the river Nene at
Peterborough relatively low flows and poor mixing between
the channel outfall and the river produce rather greater
temperature increments which are often more than 12°C

above ambient (Alabaster, 1969). As far as absolute maximum values are concerned, the Trent can reach 30°C quite frequently according to Bottomley (1971) but the highest value on record seems to be 32°C measured on the Great Ouse (Langford and Aston, 1972).

Although the highest temperatures may be short-lived and, for example, rarely exceed 3-4 hours in any day on the Severn, they can greatly modify the diurnal fluc-tuations. Thus, at Ironbridge, the daily range in summer can be doubled whilst, during winter low flows, the down-stream amplitude can reach 3-4°C with no discernible variation upstream (Langford and Aston, 1972). Ross (1970) has drawn attention to rapid fluctuations of 12-15°C in the Peterborough Cut and a diurnal variation of 8°C in early winter below a power station outfall on the river Lea enabled Gameson et al. (1959) to trace the heated effluent for more than 8 km downstream.

Reservoir Releases

The presence of a reservoir on a river is likely to lead to a reduction in the natural temperature variation down-stream for two reasons. Firstly, in view of the partial dependence of temperature on flow, the smoothed hydrograph in itself will probably produce a lower thermal range. Secondly, and more important, the temperature variations will also be suppressed when either compensation or river regulating water is drawn off at depth from a reservoir which is thermally stratified. Even with the relatively small reservoirs and temperate climate of Britain, this effect can already be detected and, with the trend towards higher dams (e.g. Clywedog on the Severn is 67 m) this impact could well increase.

During the winter an inverse thermal stratification exists with colder but less dense water near the surface. However, at this season the temperature difference between surface and bottom water cannot exceed the 4°C represented by freezing conditions near the surface and maximum density conditions at the bottom. On the other hand, in summer a well-defined and highly stable stratification often develops if the surface temperature reaches 20°C (Thompson, 1954). Warm, low-density water is now concen-trated in the shallow surface epilimnion, whose base is defined by an increasingly distinct thermocline as the temperature discrepancy between the top and bottom waters increases through the summer. Below the thermocline, exists the largely stagnant hypolimnion, which is cut off from wind action and sunlight. Here temperatures probably average around 7°C and, at more than 30 m below top water level, will hardly ever rise above 13°C.

From a study of the river Lune, a tributary of the Tees, Lavis and Smith (1972) have demonstrated the supp-ression of the natural variability of river temperatures which occurs when compensation water is withdrawn from the hypolimnion. Figure 5A shows that, on the basis of 10-day means obtained during 1968-69, inlet and outlet river temperatures differed by about 4°C during the warmest and

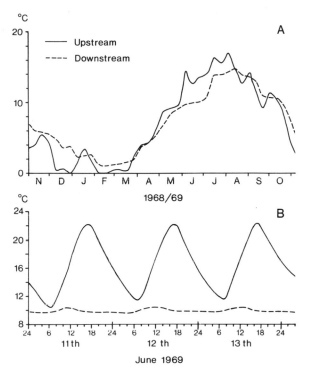

Figure 5. The effect of a direct-supply reservoir on the temperature of the river Lune.
A. 10-day means upstream and downstream from November 1968 to October 1969.
B. Continuous thermograph records for mid-June 1969. (after Lavis and Smith, 1972).

coldest parts of the year. In Figure 5B it can be seen that, in an anticyclonic period in mid-June 1969, the temperature of the compensation water, drawn off at over 28 m below top water level, remained around 10°C whilst the upstream river temperature showed a regular diurnal cycle of 10°-12°C with maximum values reaching 22°C. Therefore, immediately downstream of the reservoir, daily maxima were depressed by up to 12°C, and these summer maxima remained at least 3°C lower than the inlet temperature at a downstream distance of 3 km.

Afforestation

Any land-use changes which significantly modify local water or energy budgets are likely to have an impact on river temperatures but, in Britain, experimental evidence is so far only available for afforestation. It is well-established that, owing to the blanketing effect of the canopy on both incoming and outgoing radiation, maximum air temperatures are lower and minima are higher in the forest

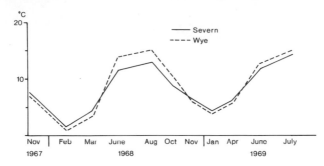

Figure 6. Mean river temperatures for the upper Severn
and Wye from November 1967 to July 1969. (after
Roberts and James, 1972).

than outside. The same applies to stream temperatures as
shown by Roberts and James (1972) in a comparison of the
open drainage basin of the river Wye with the adjacent,
coniferous-forested catchment of the upper Severn. In
Figure 6 it can be seen that, in terms of mean seasonal
values, the Severn is generally 2°C cooler in summer and up
to 1°C warmer in winter.

Similarly, the felling of woodland can re-establish the
larger range of stream temperature associated with an
unintercepted radiative flux, as reported by Gray and
Edington (1969) for a small tributary of the river Coquet.
Figure 7 illustrates the increase in diurnal fluctuation,
and the reversal in summer-temperature relationships between two sites, following the clearing of deciduous trees.
Thus, in July 1964 the warmer site B was in an open
situation 500 m upstream of forested site A. After felling
at site A, the lower station becomes warmer under
comparable weather conditions in July 1965 with a marked
increase in diurnal range leading to a maximum 6.5°C higher
than previously recorded.

Urbanization

The extension of built-up areas is also likely to interfere with the thermal regime of streams due to modifications of the local heat and moisture budgets. For
example, in the United States Pluhowski (1970) has claimed
that this process has increased average temperatures on
some Long Island streams by 5°-8°C during summer, whilst
in winter the mean values are about 1.5°-3°C lower than in
rural areas. These effects are attributed largely to
changes in short-wave radiation and groundwater conditions
but, unfortunately, no British data are available for this
specific land-use effect.

CONCLUSION

Although the thermal modification of streams and rivers is

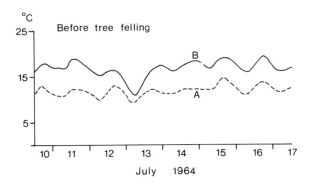

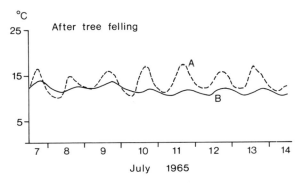

Figure 7. The change in continuous stream temperature
relationships during summer between two stations (site
A forested; site B open) following woodland clearance
around site A. (After Gray and Edington, 1969).

now widespread in Britain, it should be emphasised that
comparatively little direct ecological deterioration has so
far been attributed to such changes. Thus, according to
Langford and Aston (1972), many species of coarse fish are
capable of surviving in very changeable temperatures,
whilst fresh-water invertebrates and insects appear to
maintain essentially unchanged life-cycles at temperatures
between 28°-30°C and 26°-28°C respectively. Below Ironbridge power station there is no evidence of insects
emerging early into lethally cold air (Langford, 1975),
although on the river Trent the time of emergence of young
leeches has been brought forward by up to one month (Aston
and Brown, 1975).

On the other hand, more research is required to improve
our biological understanding of thermal modification,
especially in terms of the more indirect consequences for
fish behaviour and the effects on associated organisms in
the food chain. Wherever possible, river management should
be designed to minimise the ecological impact. For
example, power station operations could probably be related
more closely to riverflows and selective draw-off levels
could be incorporated into new high storage dams.

Finally, there is a need for a better understanding of natural river temperature behaviour on a variety of space and time scales. Thus, local physically-based studies are required to produce predictive models of river temperatures in different topographic, hydrologic and land-use situations, whilst, on a national scale, it would be useful to have a classification of river water temperatures.

19

THE UTILIZATION OF RESEARCH RESULTS
IN CATCHMENT CONTROL

O.T. Addyman

Lea Division, Thames Water Authority

ABSTRACT

*The application of research findings to catchment control
are outlined. The need to control man's actions which
have an impact on the hydrological cycle is justified in
terms of the advantages to both nature and mankind. There
is a resumé of legislation and the ways in which the water
authorities have developed a means of control.*

The control and development of man's activities as they
affect the hydrological cycle have been practised in part
probably since mankind settled in one place to live. The
effects that he has had on the environment have never been
more marked in this country than during the period from
the industrial revolution up to the present day. A lack
of understanding, the profit motive, and insufficient
control probably led more than other things to a state of
affairs where many environments in Great Britain deterior-
ated so as to become a serious threat to nature and to the
lives of the very human beings who brought about the
situation.

In the 1850s a Royal Commission commented on the quality
of the rivers Thames and Lea by suggesting that they should
not be further used for the supply of drinking water to the
population. In other parts of the country, particularly in
the industrial areas, rivers were becoming dead to nature
and unfit for human use, apart from their function as
sewers. Urban areas were developed in the flat, cheap
flood plains. Water was taken out of rivers upstream of
developments in such quantities as to completely change
their character downstream. The population of this country
began to have an awareness of what they were doing. Legislation and positive constructive action, good management
and catchment control, have created a system, recently

improved by the 1973 Water Act, which does a great deal in
the quest for the optimim for the catchment as far as man
and nature are concerned.

A great deal of what is used in control today has been
built on research foundations. As a result of the research
described in the earlier chapters of this book and elsewhere, further developments and refinements of control will
emerge. The effects of conurbations on their micro-climate
is significant and it is a sobering thought to find that
because of the rise in temperature over London the intensity
of rainfall has probably increased. An alternative theory
posed for this phenomena, namely the concentration of
condensation nuclei, led in part to thought being given,
at the time of the 1976 drought, to the use of cloud
seeding, to try and alleviate the drought in the Thames
Water area. Unfortunately any clouds that did appear were
not of the rain carrying type. Some thought ought also to
be given to the consequences of high intensity rainfall of
the type which developed at Hampstead in 1976, so that,
although flooding would result, it would only occur in
places where the damage would be minimal, with a 'fail
safe' surface water drainage system. Planning authorities
should heed water authority advice and ensure that no low
spots occur in developments which are likely to flood to
levels which cause a hazard to life.

The optimisation of land use and farming methods might
well be useful in endeavouring to produce the downstream
conditions that are desirable. Farm husbandry may in
future have to be done in such a way that the run-off from
arable land does not allow water heavy in nitrates and other
chemicals to cause problems of water quality and quantity
in the river systems. There is already criticism of the
farming industry where nitrate problems exist in river
systems and to date a lack of conclusive evidence to show
that fertilizers cause the major problem of high nitrate
values in the river systems. The research so far is useful
in assessing the situation and may in the future indicate a
need for control, although the case for this is unproven so
far.

The action of mankind on the surface soil is followed
by the action of nature in the form of wintering and
vegetation growth effects. The total effect could
no doubt lead to the adoption of ideal forms of agriculture
for a particular area. This, of course, must be one
of the future activities of our agricultural experts who
should not only know how to optimise on the crop, but should
be well versed in what its effects are likely to be on the
hydrological cycle.

The importance of the water table is probably under-
estimated from the point of view of crop production and may
be overestimated in its effects on major floods. The control of the water table by methods approaching those of
irrigation could show benefits to the catchment as far as
food production is concerned and may well be used to pres-
erve wetland areas for the conservation of the plant life
which is so important ecologically.

The effects of reservoirs on a catchment can be quite
dramatic, and nowhere better illustrates this than southern

Ireland where the electricity authority in its development of hydro-electric power earlier this century severed the salmon spawning grounds from the lower parts of the river system and reduced the flows in these sections to little more than a trickle. This very nearly destroyed the salmon industry which was an important factor in the economy of the country. Research and action followed fairly quickly after the effects became evident resulting today in a thriving fishery, an improved river system, and continuing national research into that system.

The use of reservoirs for flood control can lead to difficulties. Other uses of a reservoir are generally not compatible with that of flood control and these other uses demand that a reservoir is in a particular situation which does not give the desired flood control result. In times when a river is in flood the state of its hydrograph sometimes shows a number of peaks close together. This could mean that the first storm fills the storage available for flood water and that the second flood passes very much more rapidly across the top of the reservoir and down the catchment. This, of course, means that the catchment is probably in more danger under these circumstances than it would have been without the reservoir there. Small flood storage reservoirs, perhaps even in urban areas, can be useful in reducing peak flows downstream but it is essential to control development well into the future to ensure that the purpose of the reservoir is known and no encroachment is allowed onto the reservoir site. These areas, as in Stevenage, are often dry except when functioning as flood control reservoirs.

Flows in a channel, both in its natural state, and with its runoff altered by man's activities, generate change in the geometry of the channel by the velocity, changes of the material through which the channel flows, and the load carried by the flow from upstream. Control of any of these factors effects changes in the channel. For instance, channel revetment, a change of gradient and hence a change of velocity, or a silt trap upstream all change the effect at a particular section. In channel design these factors should be used almost as an art to the advantage of man and nature. This knowledge and the use of this understanding were demonstrated by great men of the past like Gerald Lacey, Claude Inglis and Montagu.

Water from underground has always been a significant part of mankind's supply, either taken from springs or by shallow or deep wells. It has been favoured because it has tended to be of better quality than that of natural surface sources which are more vulnerable to pollution. This attraction has caused mankind to seek and draw on the underground sources to an extent where these sources have suffered. In some areas this has caused the lowering of the ground water levels by up to 30 metres. A good example of this is the London basin, where between the 1770s and 1940s, water was abstracted in large quantities. This caused the aquifer to dry out which increased pumping heads, degraded the aquifer's properties and allowed salt water intrusion from the Thames. Research culminating in a giant pilot recharge scheme using fully treated water from the existing treatment plants at off-peak times is in hand. The information from this, together with research into management in conjunction with surface sources will produce a lead for the further development of the Thames basin and the application of the resulting techniques in other parts of the country and the world. Simulations have led to the present programme of development and will inevitably be used to solve the management problems. Total catchment studies and simulations should then enable improvements to be brought about to the catchment as a whole, possibly including the improvements of water tables and river flows at critical times.

Mankind's activites, apart from agriculture, have resulted in considerable modification of catchment behaviour. The attraction of undeveloped areas in the first place for dwellings and factory development have in more recent times caused considerable difficulty because of the scale of this development. Remaining available land is often that which was spurned in the past because the local people know of its flood liability. This fund of community experience and knowledge was developed largely because movements of population were minimal in the last century. More recently the profit motive, the limited knowledge of the past and the lack of appreciation by those concerned brought about disastrous developments in flood plains and areas subject to tidal flooding. Knowledge and a little research could have considerably reduced the need to spend the vast sums of money which have been necessary in the last 30 years. The culmination of this is London itself, where the largest single civil engineering project this country has ever known, in the form of the Thames tidal barrier, has been thought to be necessary.

Research into just what is likely to happen when developments take place is improving man's ability to control his environment. The present administration improves with each set of legislation and needs to be improved still more where the control of our rivers and flood plains are concerned.

Model experiments and simulations can help a great deal in deciding how development should take place within a catchment and there is much to be learnt from physical models rather than mathematical ones. The mathematical model means something to technical people and particularly to the individual who invented the particular model, but the value of the physical model is to indicate the impact of development to those less initiated. For example, run-off caused by a new town or development can be clearly indicated on a physical model to the members of the New Town Corporation. The need for compensatory works then becomes obvious and the way in which they may be used for other purposes can then be demonstrated to the audience. The tools of research then become the tools of management.

It is interesting to note that the major effects from urbanization as far as runoff is concerned, are more critical for the annual flood than in the longer term storms of 30 to 100 years return period.

There is a great deal yet to be found out about flood plains. They are the product of natural processes and serve as storage areas for flood water: however such a limited function is not usually the most beneficial to mankind. Probably one of the major challenges that planners and the water industry have today is to set these optimum criteria for the future. Flood plains could be modified probably quite cheaply to enable land within the present flood plain to be used for other purposes, whilst the main channel system could still convey the flood waters safely to the sea. Wetland habitats could be maintained and recreational areas of all kinds could be developed in the overall plan.

The problems of the motorway system and other road works have highlighted the need for research into the quantity and quality of runoff from these systems. The early failures on the Preston and Lancaster motorways highlighted the need for a better system of highway drainage, and for the calculation of culvert and bridge size. The frequent occurrence of flooding in towns where the TRRL method has been used indicates that further research is needed, particularly into the frequency of storm return which is used. In large urban areas it is particularly important that this should be researched once again when it is realised that all those small pockets of land which had not been used because of problems in the past are being developed because of the very great pressure on development land.

Water quality problems and their management have received considerable attention. As far as ground water is concerned past actions may well bring about critical situations in the future when polluted water, out of our rubbish dumps, re-appears in our water supply. Research on the limitation and control of these effects is to be welcomed. Possibly if the highly toxic substances are dumped in unsatisfactory conditions and henceforth cause death or permanent injury then perhaps this should be treated as though it was as serious an offence as murder.

The effects of changing the quality of water by raising its temperature is most interesting and it seems to me that the heat available in waste water should be put to more use than it is today. Perhaps research here into these possible uses might be more profitable than allowing the water to disperse its energy into the river water. It may even be suggested that the energy producing industries would be better under the unified control of the water authorities.

Water authorities exercise control and effect change in the hydrological cycle by advisory and statutory means. The advisory role is a very wide one and includes the passing of expert advice on legal, technical and financial matters associated to the following: (a) central government departments, (b) local government departments, (c) National Water Council, (d) navigation authorities, (e) planning authorities, (f) statutory undertakers, (g) other water authorities and (h) other organisations, societies, bodies and individuals. This means is used to secure or change legislation by which the water authorities apply statutory control.

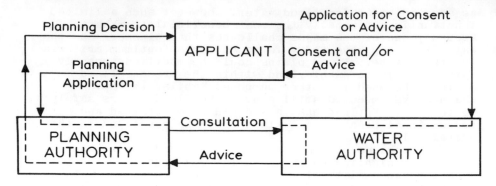

Figure 1.   The procedures used in catchment control.

The statutory function is used to control, where appropriate, changes which individuals or bodies wish to make to the present disposition of the water cycle. The legislation embodies the level and degree of control which the water authorities are able to apply in order to obtain the desired objectives and policies which they have.

A host of legislation exists enabling control in the following fields: (a) land drainage, (b) tidal defences, (c) water supply, (d) river pollution prevention, (e) water resources, (f) sewerage and sewage disposal, (g) trade effluent, (h) public health, (i) fisheries, (j) amenity and recreation. In practical terms this is achieved by the issue of consents, licences, agreements and by legal action through the courts.

In most cases at divisional level the control procedure starts either by an application by a body or individual for consent and/or advice, or by a planning application consultation from a planning authority with respect to an application by an individual or body. The first procedure often follows as a result of the second.

Figure 1 illustrates the routes for the procedures or variations of both. The control is thus effected directly by the water authority, or indirectly by the planning authority in the form of their planning decision with the appropriate conditions or agreements attached thereto.

BIBLIOGRAPHY

Ackerman, B., 1974. Wind fields over the St. Louis metropolitan area. *Journal of the Air Pollution Control Association,* 23, 232-236.

Ackermann, W.C., White, G.F. and Worthington, E.B., 1973. *Man made lakes: their problems and environmental effects.* American Geophysical Union, Geophysical Monograph 17.

Ackers, P. and Charlton, F.G., 1970. The geometry of small meandering streams. *Proceedings of the Institution Civil Engineers,* 73285, 289-317.

Affholder, M., Basalo, C., Dorfmann, R. and Fourment, P., 1973. Les décharges contrôlées de residus urbains. *Techniques et Sciences Municipales L'Eau,* 68, 273-304.

Aitken, A.P., 1975. *Hydrologic investigation and design of urban stormwater drainage systems.* Australian Water Resources Council, Technical Paper No. 10, Canberra.

Alabaster, J.S., 1969. Effects of heated discharges on freshwater fish in Britain. In: Krenkel P.A. and Parker, R.L. (Editors), *Biological aspects of thermal pollution,* Vanderbilt University Press, 354-370.

Alabaster, J.S., 1970. River flow and upstream movement and catch of migratory salmonids. *Journal of Fish Biology,* 2, 1-13.

Allan, J.F., 1969. Research needs for thermal-pollution control. In: Krenkel P.A. and Parker F.L. (Editors) *Biological aspects of thermal pollution,* Vanderbilt University Press, 382-392.

Allen, J.R.L., 1970. *Physical processes of sedimentation.* Unwin, London.

A.S.C.E., 1975. Urban sediment problems: a statement on scope, research, legislation and education. *Proceedings of the American Society Civil Engineers, Journal of the Hydraulics Division,* 101, 329-340.

Anderson, D.G., 1970. *Effects of urban development on floods in northern Virginia.* U.S. Geological Survey Water Supply Paper, 2001-C.

Anderson, J.R. and Dornbush, J.N., 1967. Influence of sanitary landfill on groundwater quality. *Journal of American Water Works Association,* 59, 457-470.

Anderson, M.G., 1975. Demodulation of stream-flow series. *Journal of Hydrology,* 26, 115-121.

Angell, J.K., Hoecker, W.H., Dickson, C.R. and Pack, D.H., 1973. Urban influences on a strong daytime air flow as determined by tetroon flights. *Journal of Applied Meteorology,* 12, 924-936.

Angino, E.E., Magnuson, L.M. and Stewart, G.F., 1972. Effects of urbanisation on stormwater runoff quality: a limited experiment, Naismith Ditch, Lawrence, Kansas. *Water Resources Research,* 8, 135-140.

Anon., 1954a. *The calculation of irrigation need.* Ministry of Agriculture Fisheries and Food. Technical Bulletin, 4, H.M.S.O., London.

Anon., 1954b. *Report on the investigation of leaching of a sanitary landfill.* State of California, Water Pollution Control Board, Publication 10.

Anon., 1961. *Pollution of water by tipped refuse.* Report of the Technical Committee on the Experimental Disposal of House Refuse in Wet and Dry Pits, Ministry Housing and Local Government, H.M.S.O., London.

Anon., 1970. *Disposal of solid toxic wastes.* Report of the Technical Committee on the Disposal of Toxic Solid Wastes, Ministry of Housing and Local Government, H.M.S.O., London.

Anon., 1971. *Groundwater resources of the Lincolnshire Limestone in South Lincolnshire.* Progress report, Welland and Nene River Authority.

Anon., 1972. *Groundwater resources of the Lincolnshire Limestone in South Lincolnshire.* Progress report, Welland and Nene River Authority.

Anon., 1973. *Groundwater resources of the Lincolnshire Limestone in South Lincolnshire.* Progress report, Welland and Nene River Authority.

Anon., 1975. *Programme of research on the behaviour of hazardous wastes in landfill sites.* Interim report on progress, Department of the Environment, London.

Anon., 1976. *Groundwater resources of the Southern Lincolnshire Limestone.* Summary of investigations, Anglian Water Authority.

Apgar, M.A. and Langmuir, D., 1971. Groundwater pollution potential of a landfill above the water table. *Groundwater,* 9, 76-96.

Appleton, B., 1975. Yorkshire Water Authority plans for cleaner rivers. *New Civil Engineer,* January 23, 1975, 21-22.

Aspinwall, R., 1974. *Pitsea project part 2: Progress in the hydrogeological evaluation of landfill sites.*

Presented at 'Waste disposal: planning the way ahead'. Conference organised by Redlands Purle/ International Systems Corporation of Lancaster, London.

Aston, R.J. and Brown, D.J.A., 1975. Local and seasonal variations in populations of the leech *Erpobdella octoculata* (L) in a polluted river warmed by condenser effluents. *Hydrobiologia,* 47, 347-366.

Atkinson, B.W., 1966. Some synoptic aspects of thunder outbreaks over south-east England 1951-60. *Weather,* 21, 203-209.

Atkinson, B.W., 1967. Structure of the thunder atmosphere, south-east England 1951-60. *Weather,* 22, 335-345.

Atkinson, B.W., 1968. A preliminary examination of the possible effect of London's urban area on the distribution of thunder rainfall 1951-60. *Transactions of the Institute of British Geographers,* 44, 97-118.

Atkinson, B.W., 1969. A further examination of the urban maximum of thunder rainfall in London 1951-60. *Transactions of the Institute of British Geographers,* 48, 97-119.

Atkinson, B.W., 1970. The reality of the urban effect on precipitation - a case study approach. In: *Urban climates,* World Meteorological Organization, Technical Note No. 108, 342-360.

Atkinson, B.W., 1971. The effect of an urban area on the precipitation from a moving thunderstorm. *Journal of Applied Meteorology,* 10, 47-55.

Atkinson, B.W., 1975. *The mechanical effect of an urban area on convective precipitation*. Department of Geography, Queen Mary College, London. Occasional Paper No. 3.

Atkinson, B.W., 1977. *Urban effects on precipitation: an investigation of London's influence on the severe storm in August 1975*. Department of Geography, Queen Mary College, London, Occasional Paper No. 8.

Avco Economic Systems Corporation, 1970. *Stormwater pollution from urban land activity*. FWQA Publication 11034, Water Pollution Control Series, Washington, D.C.

Avery, B.W., 1964. *The soils and landuse of the district around Aylesbury and Hemel Hempstead*. Sheet 238 Memoir of the Soil Survey of Great Britain, H.M.S.O. London.

Baden. W. and Eggelsmann, R., 1970. The hydrologic budget of the highbogs in the Atlantic region. In: *Proceedings of the Third International Peat Congress,* Quebec, Canada, 200-219.

Barrett, E.C., 1964. Local variations in rainfall trends in the Manchester region. *Transactions of the Institute of British Geographers,* 35, 55-72.

Battan, L.J., 1965. Some factors governing precipitation and lightning from convective cloud. *Journal of Atmospheric Science,* 22, 79-84.

Bauer, L. and Tille, W., 1967. Regional differentiation of the suspended sediment transport in Thuringia and their relation to soil erosion. In: *General assembly of Bern - river morphology,* International Association of Hydrological Sciences Publication 75, 367-377.

Bay, R.R., 1966. Evaluation of an evapotranspirometer for peat bogs. *Water Resources Research,* 2, 437-442.

Bay, R.R., 1967. Factors affecting soil moisture relationships. In: Sopper W.E. and Lull H.W. (Editors), *International symposium on forest hydrology,* Pergamon, Oxford, 335-342.

Bay, R.R., 1969. Runoff from small peat watersheds. *Journal of Hydrology,* 9, 90-102.

Beckinsale, P.O., 1972. Clearwater erosion. *International Geographer,* 2, 1244-1246.

Behilak, S.A., 1967. *Some studies in urban climatology in Sheffield in relation to air pollution.* Unpublished Ph.D. thesis, University of Sheffield.

Benoit, G.R., 1973. Effect of freeze-thaw cycles on aggregate stability and hydraulic conductivity of three soil aggregate sizes. *Proceedings of the Soil Science Society America,* 37, 3-5.

Bevan R.E., 1967. *Notes on the science and practice of controlled tipping of refuse.* Institute of Public Cleansing.

Billington, R.H. and Tester, D.J., 1973. The problems of River Authorities. In: *Proceedings of symposium on disposal of municipal and industrial sludges and solid toxic wastes,* Institute Water Pollution Control, 41-49.

Birtles, A.B., and Nutbrown, D.A., 1976. The use of groundwater modelling techniques in water resources planning. *Water Services,* 80, 533-538.

Bisal, F. and Nielsen, K.F., 1964. Soil aggregates do not necessarily break down over winter. *Soil Science,* 98, 345-346.

Black, J.H., Boreham, D., Bromley, J., Campbell, D.J.V., Mather, J.D. and Parker, A. 1976. *Construction and instrumentation of lysimeters to study pollutant movement through unsaturated sand.* Paper presented at Water Research Centre symposium on groundwater quality, measurement, prediction and protection. Reading, 6-8 September, 1976.

Blyth, K. and Kidd, C.H.R., 1977. The development of a meter for the measurement of discharge through a road gully. *Chartered Municipal Engineer,* February 1977.

Boelter, D.H., 1964. Water storage characteristics of several peats. *Proceedings of the Soil Science Society of America,* 28, 433-435.

Boelter, D.H., 1969. Physical properties of peat as related to degree of decomposition. *Proceedings of the Soil Science Society of America,* 33, 606-609.

Boelter, D.H., 1972. Water-table drawdown around an open ditch in organic soil. *Journal of Hydrology,* 15, 329-340.

Boelter, D.H., 1975. Hydraulic conductivity of peats. *Soil Science,* 100, 227-230.

Boon, P.J. and Shires, S.W., 1976. Temperature studies on a river system in north-east England. *Freshwater Biology,* 6, 23-32.

Bornstein, R.D., Lorenze, A. and Johnson, D., 1972. Recent observations of urban effects on winds and temperatures in and around New York City. In: *Preprints conference on urban environment and second conference on biometeorology,* American Meteorological Society, Boston, 28-33.

Bottomley, P., 1971. Contribution to *Symposium on Freshwater Biology and Electrical Power Generation,* 22 April, Central Electricity Research Laboratories, Leatherhead.

Bow, C.J., Howell, F.T. and Thompson, P.J., 1969. The lowering of the water table in the Permo-Triassic rocks of South Lancashire. *Water and Water Engineering,* 73, 461-463.

Braham, R.R. and Spyers-Duran, P.A., 1974. Ice nucleus measurements in an urban atmosphere. *Journal of Applied Meteorology,* 13, 940-945.

Brandt, G.H., 1972. *An economic analysis of erosion and sediment control methods for watersheds undergoing urbanization.* Dow Chemical Co. Report, Midland, Michigan.

Brown, J., 1882. *The forester,* 5th edition. Wm Blackwood and Sons, Edinburgh and London.

Brown, R.M., 1975. *Chemical control of weeds in the forest.* Forestry Commission Booklet 40, H.M.S.O., London.

Bryan, E.H., 1970. *Quality of stormwater drainage from urban land.* Water Resources Research Institute Univ. of N. Carolina, Report 37, Durham.

Burke, W., 1969. Drainage in blanket peat at Glenamoy. In: *Transactions of the Second International Peat Congress,* Leningrad, USSR, 809-817.

Burke, W., 1975. Aspects of the hydrology of blanket peat in Ireland. In: *Hydrology of marsh-ridden areas,* Studies and Reports in Hydrology 19, UNESCO - International Association of Hydrological Sciences. 171-182.

Burwell, R.E., Allmaras, R.R. and Amemiya, A., 1963. A field measurement of total porosity and surface microrelief of soils. *Proceedings of the Soil Science Society of America,* 27, 697-700.

Cairns, J., 1967. The use of quality-control techniques in the management of aquatic ecosystems. *Water Resources Bulletin,* 3, 47-53.

Carter, R.W., 1961. Magnitude and frequency of floods in suburban areas. *U.S. Geological Survey Professional Paper,* 424-B, 9-11.

Chandler, T.J., 1965. *The climate of London,* Hutchinson.

Chandler, T.J., 1967. Absolute and relative humidities in towns. *Bulletin of the American Meteorological Society,* 48, 394-399.

Changnon, S.A., 1961. A climatological evaluation of precipitation patterns over an urban area. In: *Air over cities,* U.S. Dept. Health, Education and Welfare, Public Health Service, 37-67.

Changnon, S.A., 1976. Inadvertent weather modification. *Water Resources Bulletin,* 12, 695-718.

Changnon, S.A., Huff, F.A. and Semonin, R.G., 1971. METROMEX: an investigation of inadvertent weather modification. *Bulletin of the American Meteorological Society,* 52, 958-967.

Chapman, S.B., 1965. The ecology of Coom Rig Moss, Northumberland. *Journal of Ecology,* 53, 371-384.

Chen, C.N., 1974. Effect of land development on soil erosion and sediment concentration in an urbanising basin. In: *Effects of man on the interface of the hydrological cycle with the physical environment,* International Association of Hydrological Sciences Pub. 113, 150-157.

Chow, V.T., 1964. Chapter 14, Runoff. In: Chow V.T. (Editor), *Handbook of applied hydrology,* McGraw-Hill, New York.

Clark, C.O., 1945. Storage and the unit hydrograph. *Transactions of the American Society of Civil Engineers,* 110, 1419-1446.

Clark, D. and England, G., 1963. Thermal power generation. In: *Conservation of water resources in the U.K.* Institution Civil Engineers, London, 43-51.

Clarke, R.T. and Newson, M.D., (in press). Some detailed water balance studies of research catchments. *Philosophical transactions of the Royal Society, London.*

Clayton, K.M. and Brown, J.C., 1958. The glacial deposits around Hertford. *Proceedings of the Geologists Association,* 69(2), 103-119.

Cohen, P., and Franke, O.L. and Foxworthy, B.L., 1968. *An atlas of Long Island's water resources.* New York State Water Resources Commission, Bulletin 62, Albany, N.Y.

Cole, G., 1976. Land drainage in England and Wales. *Journal of Institution Water Engineers and Scientists,* 30, 345-367.

Colston, N.V., 1971. *Characteristics and treatment of urban land runoff.* Environmental Protection Agency, Report 670/2-74-096, Washington, D.C.

Common, R., 1970. Land drainage and water use in Ireland. In: Stephens N. and Glasscock R.E. (Editors), *Irish Geographical Studies,* Queen's University, Belfast, 342-359.

Construction Industry Research and Information Association (CIRIA), 1974. *Rainfall, runoff and surface water drainage of urban catchments.* Proceedings of the 1973 Bristol Symposium, CIRIA, London.

Conway, V.M. and Miller, A., 1960. The hydrology of some small peat-covered catchments in the northern Pennines. *Journal of the Institution Water Engineers,* 14, 415-424.

Cordery, I., 1977. Quality characteristics of urban stormwater in Sydney, Australia. *Water Resources Research,* 13, 197-202.

Corish, P., 1971. *Effects of arterial drainage on low flow in the Nenagh River* (Mimeo).

Costin, A.B. and Dooge, J.C.L., 1973. Balancing the effects of man's actions on the hydrological cycle. In: *Man's influence on the hydrological cycle,* FAO Irrigation and Drainage Paper 17, 19-52.

Crippen, J.R., 1965. Changes in character of unit hydrographs, Sharon Creek, California, after suburban development, *U.S. Geological Survey Professional Paper* 525-D, 196-198.

Crisp, D.T. and Le Cren, E.D., 1970. The temperature of three different small streams in north-west England. *Hydrobiologia,* 35, 305-323.

Davies, E.G., 1944. Ffigyn Blaen Brefi: a Welsh upland bog. *Journal of Ecology,* 32, 147-166.

Dawdy, D.R., 1967. Knowledge of sedimentation in urban environments. *Proceedings of the American Society of Civil Engineers, Journal of the Hydraulics Division,* 93 HY6, 235-245.

Dawdt, D.R., Lichty, R.W. and Bergmann, J.M., 1972. A rainfall-runoff simulation model for estimation of flood peaks for small drainage basins.

U.S. Geological Survey Professional Paper 506-B, 1-28.

De Coursey, D.G., 1975. Implications of floodwater retarding structures. *Transactions of the American Society Agricultural Engineers,* 18(5), 897-904.

De Filippi, J.A. and Shih, C.S., 1971. Characteristics of separated storm and combined sewer flows. *Water Pollution Control Federation,* 43, 85-92.

Deij, L.J.L., 1956. Contributions to a discussion on evaporation. *Netherlands Journal of Agricultural Science,* 4, 92.

Delfs, J., 1967. Interception and stemflow in stands of Norway spruce and Beech in West Germany. In: Sopper, W.E. and Lull, W.H. (Editors), *International symposium on forest hydrology,* Pergamon, Oxford, 179-183.

Dempster, G.R., 1974. *Effects of urbanisation on floods in the Dallas, Texas, metropolitan area.* U.S. Geological Survey Water Resources Investigations 60-73, Austin, Texas.

Detwyler, T.R., 1971. *Man's impact on environment.* McGraw Hill, New York.

Dooge, J., 1975. The water balance of bogs and fens. *Hydrology of marsh ridden areas,* Studies and Reports in Hydrology No. 19, UNESCO - International Association Hydrological Sciences, 233-271.

Downing, R.A., and Oakes, D.B., Wilkinson, W.B. and Wright, C.E., 1974. Regional development of groundwater resources in combination with surface water. *Journal of Hydrology,* 22, 155-177.

Downing, R.A., Smith, D.B., Pearson, D.J., Monkhouse, R.A. and Otler, R.L., 1977. The age of groundwater in the Lincolnshire Limestone, England and its relevance to the flow mechanism. *Journal of Hydrology,* 33, 201-216.

Downing, R.A., and Williams, B.P.J., 1969. *The groundwater hydrology of the Lincolnshire Limestone with special reference to groundwater resources.* Water Resources Board Publication 9, H.M.S.O., London.

Durbin, T.V., 1974. *Digital simulation of the effects of urbanisation on runoff in the upper Santa Ana valley, California,* U.S. Geological Survey Water Resource Investigations, 41-73, Menlo Park, California.

Edington, J.M., 1966. Some observations on stream temperature, *Oikos,* 15, 265-273.

Edwards, K.E. and Rodda, J.C., 1972. Preliminary study of the water balance of a small clay catchment. In: *The results of research on representative and experimental basins,* International Association of

Hydrological Sciences Publication 97, 187-199.

Ellis, J.B., 1975. *Urban stormwater pollution*. Middlesex Polytechnic, Hendon, Research Report 1, (Mimeo).

Ellis, J.B., 1976. Sediments and water quality of urban stormwater. *Water Services*, 80, 730-734.

Eggelsmann, R., 1975. The water-balance of lowland areas in North-western regions of the FRG. In: *Hydrology of marsh-ridden areas*, Studies and reports in hydrology, 19, UNESCO - International Association of Hydrological Sciences, 355-367.

Ellison, W.D., 1947a. Soil erosion studies II: soil detachment hazard by raindrop splash. *Agricultural Engineering*, 28, 197-201.

Ellison, W.D., 1947b. Soil erosion studies III: some effects of soil erosion in infiltration and surface runoff. *Agricultural Engineering*, 28, 245-248.

Emmett, W.W., 1974. Channel changes. *Geology*, 2, 271-272.

Espey, W.H., Jr., Morgan, C.W., and Masch, F.D., 1965. *A study of some effects of urbanisation on storm runoff from a small watershed*. Texas University, Centre for Research in Water Resources, Hydraulic Engineering Laboratory. Report HYD-07-6501-CRWR-2, Austin, Texas.

Espey, W.H., Jr., Winslow, D.E. and Morgan, C.W., 1969. Urban effects on unit hydrographs. In: Moore, W.L. and Morgan, C.W., (Editors), *Effects of watershed changes on streamflow*, University of Texas Press, Austin, 215-228.

Evelyn, J.B., Narayana, V.V.D., Riley, J.P. and Israelsen, E.K., 1970. *Hydrograph synthesis for watershed subzones from measured urban parameters*. Utah State University, Utah Water Research Laboratory Report, Logan, Utah.

FAO, 1973. *Man's influence on the hydrological cycle*. FAO Irrigation and Drainage Paper 17.

Farquhar, G.J., Farvolden, R.N., Hill, H.M. and Rovers, F.A., 1972. *Sanitary landfill study final report, Vol. 1*. Waterloo Research Institute.

Farquhar, G.J. and Rovers, F.A., 1975. *Guidelines to landfill location and management for water pollution control*. Sanitary Landfill Study 4, Waterloo Research Institute.

Farquhar, G.J. and Rovers, F.A., 1976. Leachate attenuation in undisturbed and remoulded soils. In: *Gas and leachate from landfills: formation, collection and treatment*, U.S. Environmental Protection Agency, Report 600/9-76-004, 54-70.

Fitzgerald, J.W. and Spyers-Duran, P.A., 1973. Changes in cloud nucleus concentration and cloud droplet size distribution associated with pollution from St. Louis. *Journal of Applied Meteorology*, 12, 511-516.

Forestry Commission, 1928. *Report on census of woodlands and census of production of home-grown timber, 1924.* H.M.S.O., London.

Forestry Commission, 1976. *56th annual report and accounts 1975-76.* H.M.S.O., London.

Forster, V. and Muller, G., 1973. Heavy metal accumulation in river sediments. *Geoforum*, 14, 53-61.

Foster, S.S.D. and Crease, R.I., 1974. Nitrate pollution of chalk groundwater in East Yorkshire - a hydro-geological appraisal. *Journal of the Institution of Water Engineers and Scientists*, 28, 178-194.

Fourt, D.F. and Hinson, W.H., 1970. Water relations of tree crops: a comparison between Corsican pine and Douglas fir in south-east England. *Journal of Applied Ecology*, 7(2), 295-309.

Fox, I.A. and Rushton, K.R., 1976. Rapid recharge in a limestone aquifer. *Groundwater*, 14, 21-27.

Fuller, W.H. and Korte, N., 1976. Attenuation mechanisms of pollutants through soils. In: *Gas and leachate from landfills formation, collection and treatment*, U.S. Environmental Protection Agency, Report 600/9-76-004, 111-121.

Fungaroli, A.A. and Steiner, R.L., 1971. Laboratory study of the behaviour of a sanitary landfill. *Journal of the Water Pollution Control Federation*, 43, 252-67.

Gameson, A.L.H., Gibbs, J.W. and Barrett, M.J., 1959. A preliminary temperature survey of a heated river. *Water and Water Engineering*, 63, 13-17.

Gameson, A.L.H., Hall, H. and Preddy, W.S., 1957. Effects of heated discharges on the temperature of the Thames estuary. *Engineer*, 204, 3-12.

Gardiner, J., 1974. Chemistry of cadmium in natural water. Part II: The absorption of cadmium on river muds and naturally occuring solids. *Water Resources*, 8, 157-164.

Geswein, A.J., 1975. *Liners for land disposal sites; an assessment*. U.S. Environmental Protection Agency Report SW-137.

Gilbert, C.R. and Satier, S.P., 1970. *Hydrologic effects of floodwater retarding structures, Garza-Little Elm reservoirs, Texas, U.S.* U.S. Geological Survey Water Supply Paper, 1984.

Glymph, L.M. and Holtan, H.N., 1969. Land treatment in agricultural watershed hydrology research. In:

Moore, W.L. and Morgan, C.W., (Editors), *Effects of watershed changes on streamflow*. University of Texas Press, Austin, 44-68.

Golwer, A. and Matthess, G., 1968. Research on groundwater contaminated by deposits of solid waste. *Bulletin of the International Association of Scientific Hydrology,* 78, 129-33.

Graf, W.H., 1971. *Hydraulics of sediment transport.* McGraw-Hill, New York.

Graf, W.L., 1975. The impact of suburbanisation on fluvial geomorphology. *Water Resources Research,* 11, 690-692.

Gray, D.A., Mather, J.D. and Harrison, I.B., 1974. Review of groundwater pollution from waste disposal sites in England and Wales, with provisional guidelines for future site selection. *Quarterly Journal of Engineering Geology,* 7, 181-96.

Gray, J.R.A. and Edington, J.M., 1969. Effect of woodland clearance on stream temperature. *Journal of the Fisheries Research Board of Canada,* 26(2), 399-403.

Gray, R. and Henton, M.P., 1975. An investigation into the contamination of groundwater at an industrial tipping site. *Solid Wastes,* 65, 153-170.

Green, E.H., 1974. *Black deposits on motorways.* Transport and Road Research Laboratory Supplementary Report 74 UC, Crowthorne, Berks.

Green, F.H.W., 1959. Some observations of potential evapo-transpiration, 1955-57. *Quarterly Journal of the Royal Meteorological Society,* 85, 152-158.

Green, F.H.W., 1973. Aspects of the changing environment: some factors affecting the aquatic environment in recent years. *Journal of Environmental Management,* 1, 377-391.

Green, F.H.W., 1974. Changes in artificial drainage, fertilizers, and climate in Scotland. *Journal of Environmental Management,* 2, 107-122.

Green, F.H.W., 1975. The effect of climatic and other environmental changes on water quality in rural areas. In: Hey, R.D. and Davis, T.D. (Editors), *Science, technology and environmental management,* Saxon House, 123-135.

Green, F.H.W., 1976. Recent changes in land use and treatment. *Geographical Journal,* 142, 12-26.

Greenland, D.J., Rimmer, D. and Payne, D., 1975. Determination of the structural stability class of English and Welsh soils using a water coherence test. *Journal of Soil Science,* 26, 294-303.

Gregory, K.J., 1974. Streamflow and building activity. In: Gregory, K.J. and Walling, D.E. (Editors), *Fluvial processes in instrumented watersheds.*

Institute British Geographers Special Publication
6, 169-191.

Gregory, K.J., 1976. Lichens and the determination of
river channel capacity. *Earth Surface Processes,*
1, 273-285.

Gregory, K.J. and Park, C.C., 1974. Adjustment of river
channel capacity downstream from a reservoir.
Water Resources Research, 10, 870-877.

Gregory, K.J. and Park, C.C. 1976a. Stream channel morphology in northwest Yorkshire. *Revue de Geomorpho-
logie Dynamique,* 2, 63-72.

Gregory, K.J. and Park, C.C., 1976b. The development of
a Devon gully and man. *Geography,* 61, 77-82.

Gregory, K.J. and Walling, D.E., 1973. *Drainage basin form
and process.* Arnold, London.

Griffin, R.A., Cartwright, K., Shimp, N.F., Steele, J.D.,
Ruch, R.R., White, W.A., Hughes, G.M. and
Gilkeson, R.H., 1976. *Attenuation of pollutants in
municipal landfill leachate by clay minerals.
Part 1 - Column leaching and field verification.*
Environmental Geology Notes. Illinois Geological
Survey, 78.

Griffin, R.A., Frost, R.R., Au, A.K., Robinson, G.D. and
Shimp, N.F., 1977. *Attenuation of pollutants in
municipal landfill leachate by clay minerals.
Part 2 - Heavy-metal adsorption.* Environmental
Geology Notes, Illinois Geological Survey, 79.

Grindley, J., 1970. Estimation and mapping of evaporation.
In: *Proceedings symposium on world water balance,*
Vol. 1. International Association of Scientific
Hydrology Publication 92, 200-213.

Grimshaw, D.L. and Lewin, J., in preparation. Reservoir
effects on sediment yield.

Gunn, R. and Phillips, B.B., 1957. An experimental investigation of the effect of air pollution on the
initiation of rain. *Journal of Meteorology,* 14,
272-280.

Guy, H.P., 1965. Residential construction and sedimentation
at Kensington, Md. In: *U.S. Department of Agriculture, Agricultural Research Service Miscellaneous
Publication* 970, 30-37.

Guy, H.P., 1970. *Sediment problems in urban areas.* U.S.
Geological Survey Circular 601-E.

Guy, H.P., 1972. Urban sedimentation in perspective.
*Proceedings of the American Society of Civil
Engineers, Journal of the Hydraulics Division,* 98,
209-216.

Haith, D.A., 1976. Land use and water quality in New York
rivers. *American Society of Civil Engineers,
Journal Environmental Engineering,* EEI, 1-15.

Hall, D.G., 1967. The pattern of sediment movement in the River Tyne. *International Association Scientific Hydrology Publication* 75, 117-142.

Hall, M.J., 1973. Synthetic unit hydrograph technique for the design of flood alleviation works in urban areas. In: *The design of water resources projects with inadequate data,* UNESCO/WMO/IAHS Symposium, Vol.1, 145-161.

Hammer, T.R., 1973. *Effects of urbanisation on stream channels and stream flow.* Regional Science Research Institute, Philadelphia, Pa.

Harley, B.M., Perkins, F.E. and Eagleson, P.S., 1970. *A modular distributed model of catchment dynamics.* Massachusetts Institute of Technology, Ralph M. Parsons Laboratory, Report No.133.

Harvey, A.M., 1969. Channel capacity and the adjustment of streams to hydrologic regime. *Journal of Hydrology,* 8, 82-98.

Hathaway, C.A., 1948. Observations of channel changes, degradation and scour below dams. In: *Report of 2nd Congress of the International Association of Hydrological Research,* Stockholm, 267-307.

Hawes, F.B., 1970. Thermal problems 'old hat' in Britain. *Electrical World,* April, 40-42.

Hawes, F.B., 1971. Water in electricity generation. In: *Symposium on freshwater biology and electrical power generation,* 22 April, Central Electricity Research Laboratories, Leatherhead.

Headworth, H.G., 1970. The selection of root constants for the calculation of actual evaporation and infiltration for chalk catchments. *Journal of the Institution of Water Engineers,* 24, 431-446.

Headworth, H.G., 1972. The analysis of natural groundwater level fluctuations in the Chalk of Hampshire. *Journal of the Institution of Water Engineers,* 26, 107-124.

Hedley, G., and Lockley, J.C., 1975. Quality of water discharged from an urban motorway. *Water Pollution Control,* 74, 659-674.

Helliwell, P.R., 1978. *Urban storm drainage.* Pentech Press Plymouth.

Herschy, R.W., 1965. River water temperature. *Water Resources Board Technical Note* No. 5.

Hey, R.D., 1974. Prediction and effects of flooding in alluvial systems. In: Funnell, B.M. (Editor), *Prediction in geological hazards.* Geological Society Miscellaneous Paper No.3, 42-56.

Hey, R.D., 1975a. Design discharges for natural channels. In: Hey, R.D. and Davies, T.D. (Editors), *Science, technology and environmental management,* Saxon House, D.G. Heath Ltd., U.K.

Hey, R.D., 1975b.   Response of alluvial channels to river
    regulation. In: *Proceedings of the 2nd World
    Congress, International Water Research Association,*
    New Delhi, Vol 5, 183-188.

Hibbert, A.R., 1967.   Forest treatment effects on water
    yield. In: Sopper, W.E. and Lull, H.W. (Editors),
    International symposium on forest hydrology,
    Pergamon, Oxford, 527-544.

Hibbert, A.R. and Cunningham, G.B., 1967.   Streamflow data
    processing opportunities and application. In:
    Sopper, W.E. and Lull, H.W. (Editors), *Inter-
    national symposium on forest hydrology,* Pergamon,
    Oxford, 725-736.

Hill, A.R., 1976.   The environmental impact of agricultural
    land drainage. *Journal of Environmental Management,*
    4, 251-274.

Hindley, R., 1965.   Sink holes on the Lincolnshire Lime-
    stone between Grantham and Stamford. *East Midlands
    Geographer,* 3, 454-460.

Hobbs, P.V., Radke, L.F. and Shumway, S.E., 1970.   Cloud
    condensation nuclei from industrial sources and
    their apparent influence on precipitation in
    Washington State. *Journal of Atmospheric Science,*
    27, 81-89.

Holeman, J.N. and Geiger, A.F., 1965.   Sedimentation of
    Loch Raven and Prettyboy Reservoirs, Baltimore
    County, Maryland. *U.S. Department of Agriculture,
    Soil Conservation Service Technical Paper* 136.

Hollis, G.E., 1974.   The effect of urbanisation on floods
    in the Canon's Brook, Harlow, Essex. In: Gregory,
    K.J. and Walling, D.E. (Editors), *Fluvial processes
    in instrumented watersheds,* Institute British
    Geographers, Special Publication 6, 123-139.

Hollis, G.E., 1975.   The effect of urbanisation on floods
    of different recurrence intervals. *Water Resources
    Research,* 11, 431-434.

Hollis, G.E. and Luckett, J.K., 1976.   The response of
    natural river channels to urbanisation: two case
    studies from southeast England. *Journal of Hydro-
    logy,* 30, 351-363.

Holden, A.V., 1976.   The relative importance of agri-
    cultural fertilisers as a source of nitrogen and
    phosphorus in Loch Leven. Ministry of Agriculture,
    Fisheries and Food Technical Bulletin No. 32
    In: *Agriculture and Water Quality,*   306-314.

Holstener-Jørgensen, H., 1967.   Influences of forest
    management and drainage on ground water fluctua-
    tions. In: Sopper, W.E. and Lull, H.W. (Editors),
    International symposium on forest hydrology,
    Pergamon, Oxford, 325-332.

Hopkins, G.J. and Popalisky, J.R., 1970.   Influence of an
    industrial waste landfill operation on a public

water supply. *Journal of the Water Pollution Control Federation,* 42, 431-436.

Horton, R.E., 1936. Hydrologic interrelations of water and soils. *Proceedings of the Soil Science Society of America,* 1, 401-429.

Houle, M.J., Bell, R.E., Long, D.E., Soyland, J.E. and Roulier, M., 1976. Industrial hazardous waste migration potential. In: *Residual management by land disposal.* U.S. Environmental Protection Agency, Report No. 600/9-76-015, 76-85.

Hudson, N.W., 1957. Erosion control research. *Rhodesia Agricultural Journal,* 54, 297-323.

Hudson, N.W., 1971. *Soil conservation.* Batsford, London.

Huggins, A.F., and Griek, M.R., 1974. River regulation as influence on peak discharge. *Proceedings of the American Society Civil Engineers, Journal of the Hydraulics Division,* 100, HY7, 901-918.

Hughes, G.M., Landon, R.A. and Farvolden, R.N., 1971. *Hydrogeology of solid waste disposal sites in north-eastern Illinois.* U.S. Environmental Protection Agency Report SW-12d.

Hutchinson, P., 1970. The accuracy of estimates of areal mean rainfall. In: *Research on representative and experimental basins.* International Association of Scientific Hydrology-UNESCO studies and reports in hydrology, 12, 203-218.

Iles, R.B., 1963. Cultivating fish for food and sport in power station water. *New Scientist,* 17, 227-229.

Ineson, J., 1966. *The assessment of aquifer yield.* Hydrological Group of Institution of Civil Engineers, 28th April 1966. Mimeo.

Ineson, J., 1970. Development of groundwater resources in England and Wales. *Journal of the Institution of Water Engineers,* 24, 155-177.

Ineson, J. and Downing, R.A., 1964. The groundwater component of river discharge and its relationship to hydrogeology. *Journal of the Institution of Water Engineers,* 18, 519-541.

Ingram, H.A., 1967. Problems of hydrology and plant distribution in mires. *Journal of Ecology,* 55, 711-24.

Institute of Hydrology, 1973. *Research 1972-73.*

Institute of Hydrology, 1976. *Water balance of the head-water catchments of the Wye and Severn, 1970-75,* Institute of Hydrology Report No. 33.

Institution of Civil Engineers (I.C.E.), 1972. *Determination of residual flows in rivers.* Discussion Paper. The Institution of Civil Engineers, London.

Institution of Civil Engineers (I.C.E.), 1975a. *Reservoir flood standards.* Discussion Paper. The Institution of Civil Engineers, London.

Institution of Civil Engineers (I.C.E.), 1975b. *Engineering hydrology today*. Institution of Civil Engineers, London.

International Association of Hydrological Sciences (I.A.H.S.), 1974. *Effects of man on the interface of the hydrological cycle with the physical environment*. International Association of Hydrological Sciences Publication 113.

International Association of Hydrological Sciences (I.A.H.S.) 1977. *Effects of urbanization and industrialization on the hydrological regime and water quality*. International Association of Hydrological Sciences Publication 123.

International Hydrological Programme (I.H.P.), 1975. Scientific projects of I.H.P. *Nature and Resources*, 11(3), 25-28.

International Hydrological Programme (I.H.P.), 1978. Urban water management. *Nature and Resources*, 14(1), 24-26.

Isler & Co., 1893. Artesian bored tube well at Bourne, Lincolnshire. *Engineering*, 24, 649.

Jamieson, D.G., Radford, P.J. and Sexton, J.R., 1974. The hydrological design of water resource systems. *Water Resources Board*, Reading, England.

James, L.D., 1965. Using a digital computer to estimate the effects of urban development on flood peaks. *Water Resources Research*, 1, 223-234.

James, L.D., 1972. Hydrologic modelling, parameter estimation, and watershed characteristics. *Journal of Hydrology*, 17, 283-307.

Kidd, C.H.R., 1976. *A nonlinear urban runoff model*. Institute of Hydrology Report No.31.

Kidd, C.H.R. and Helliwell, P.R., 1977. Simulation of the inlet hydrograph for urban catchments. *Journal of Hydrology*, 35, 159-172.

Kittredge, J., 1948. *Forest influences*. McGraw-Hill, New York and London.

Kockmond, W.C. and Mack, G.J., 1972. The vertical distribution of cloud and Aitken nuclei downwind of urban pollution sources. *Journal of Applied Meteorology*, 11, 141-148.

Komura, S. and Simons, D.G., 1967. River bed degradation below dams. *Proceedings of the American Society Civil Engineers, Journal of the Hydraulics Division*, 474, 1-14.

Kopec, R.J., 1973. Daily spatial and secular variations of atmospheric humidity in a small city. *Journal of Applied Meteorology*, 12, 639-648.

Kramer, P.J., 1969. *Plant and soil-water relationships*. McGraw-Hill, New York.

Kuipers, H., 1957. A relief meter for soil cultivation

studies. *Netherlands Journal of Agricultural Research,* 5, 255-262.

Kuipers, H. and Van Oewerkerk, C., 1963. Total pore-space estimations in freshly ploughed soil. *Netherlands Journal of Agricultural Research,* 11, 45-53.

Laboratory for Computer Graphics and Spatial Analysis, 1975a. *SYMAP, Version 5.17.* Graduate School of Design, Harvard University, Cambridge, Mass.

Laboratory for Computer Graphics and Spatial Analysis, 1975b. *SYMVU.* Graduate School of Design, Harvard University, Cambridge, Mass.

Landsberg, H.E., 1956. Climate of towns. In: Thomas, W.L. (Editor), *Man's role in changing the face of the earth,* Chicago, 584-603.

Langford, K.J. and Turner, A.K., 1973. An experimental study of the application of kinematic-wave theory to overland flow. *Journal of Hydrology,* 18, 125-145.

Langford, T.E., 1970. The temperature of a British river upstream and downstream of a heated discharge from a power station. *Hydrobiologia,* 35, 353-375.

Langford, T.E., 1971. The distribution, abundance and life-histories of stoneflies (Plecoptera) and mayflies (Ephemeroptera) in a British river, warmed by cooling-water from a power station. *Hydrobiologia,* 38, 339-375.

Langford, T.E., 1972. A comparative assessment of thermal effects in some British and North American rivers. In: Oglesby, R.T. *et al.* (Editors), *River ecology and man,* Academic Press, New York and London, 319-351.

Langford, T.E., 1975. The emergence of insects from a British river warmed by power station cooling-water. Part 2. *Hydrobiologia,* 47, 91-133.

Langford, T.E. and Aston, R.J., 1972. The ecology of some British rivers in relation to warm water discharges from power stations. *Proceedings of the Royal Society London, Series B,* 180, 407-419.

Langford, T.E. and Daffern, J.R., 1975. The emergence of insects from a British river warmed by power station cooling-water. Part 1. *Hydrobiologia,* 46, 71-114.

Lanyon, R.F. and Jackson, J.P., 1974. *A streamflow model for metropolitan planning and design.* American Society Civil Engineers, Water Resources Research Program, Technical Memorandum TM20.

Laurenson, E.M., 1964. A catchment storage model for run-off routing. *Journal of Hydrology,* 2, 141-163.

Laurie, M.V., 1956. The effect of forests in water catchment areas on the water losses by evaporation and transpiration. In: *Report of 12th IUFRO Congress,*

Oxford, Vol.1, Section 11, 82-85.

Lauterbach, D. and Leder, A., 1969. The influence of reservoir storage on statistical peak flows. In: *Floods and their computation,* IASH/UNESCO/WMO. Vol. 2, 821-826.

Lavis, M.E. and Smith, K., 1972. Reservoir storage and the thermal regime of rivers, with special reference to the river Lune, Yorkshire. *Science of the Total Environment,* 1, 81-90.

Law, F., 1956. The effect of afforestation upon the yield of water catchment areas. *Journal of the British Waterworks Association,* November 1956, 489-494.

Law, F., 1957a. The effect of afforestation upon the yield of water catchment areas. *Journal of the Institution of Water Engineers,* 11, 269-276.

Law, F., 1957b. Measurement of rainfall, interception and evaporation losses in a plantation of Sitka spruce trees. In: *International Association Scientific Hydrology, General Assembly of Toronto,* Vol.2, 397-411.

Lawrence, A.R., Lloyd, J.W. and Marsh, J.M., 1976. Hydrochemistry and groundwater mixing in part of the Lincolnshire Limestone aquifer. *Groundwater,* 14, 320-327.

Leopold, L.B., 1968. *Hydrology for urban land planning - a guidebook on the hydrologic effects of urban land use.* U.S. Geological Survey Circular 554.

Leopold, L.B., 1973. River channel change with time: an example. *Bulletin of the American Geological Society,* 84, 1845-1860.

Leopold, L.B. and Maddock, T. Jr., 1954. *The flood control controversy.* Ronald, New York.

Lester, W.F., 1967. Pollution in the River Trent and its tributaries, and related problems of regeneration. *Journal of the Institution of Water Engineers,* 21, 261-274.

Lewin, J. and Brindle, B., 1977. Confined meanders. In: Gregory K.J. (Editor), *River channel changes,* Wiley, London, 221-233.

Lewin, J. and Hughes, D., in preparation. Welsh floodplain studies II: application of a qualitative inundation model.

Lewin, J. and Manton, M.M.M., 1975. Welsh floodplain studies: the nature of floodplain geometry. *Journal of Hydrology,* 25, 37-50.

Leyton, L. and Rousseau, L.Z., 1958. The relationship between the growth and mineral nutrition of conifers. In: Thimann, K.V. (Editor), *Physiology of forest trees.* Ronald, New York, 323-346.

Lieber, M., Perlmutter, N.M. and Frauenthal, H.L., 1964.

Cadmium and chromium in Nassau County groundwater. *Journal of the American Waterworks Association,* 56, 739-47.

Ligon, J.T. and Stafford, D.B., 1974. *Correlation of hydrologic model parameters with changing land use as determined from aerial photographs.* Clemson University, Department of Agricultural Engineering, South Carolina Water Resources Research Institute Report 50.

Low, A.J., 1972. The effect of cultivation on the structure and other physical characteristics of grassland and arable soils. (1945-1970). *Journal of Soil Science,* 23, 363-380.

Low, A.J., 1973. Soil structure and crop yield. *Journal of Soil Science,* 24, 249-259.

Low, A.J., 1975. *Production and use of tubed seedlings.* Forestry Commission Bulletin 53, H.M.S.O., London.

Low, A.J. and Stuart, P.R., 1974. Micro-structural differences between arable and old grassland soils as shown in the scanning electron microscope. *Journal of Soil Science,* 25, 135-137.

Lowing, M.J., 1976. *Urban hydrological modelling and catchment research in the U.K.* Contribution to I.H.P. Projects 7.1, 7.2, UNESCO, Amsterdam, (Mimeo).

Lowing, M.J., 1977. *Urban hydrological modelling and catchment research in the U.K.* Institute of Hydrology Report No.36.

Macan, T.T., 1958. The temperature of a small stony stream. *Hydrobiologia,* 12, 89-106.

Macan, T.T., 1960. The occurrence of *Heptagenia lateralis* (Ephem.) in streams in the English Lake District. *Wetter und Leben,* 12, 231-234.

Macan, T.T., 1974. *Freshwater ecology.* (2nd Ed.) Longmans, London.

Marsalek, J., 1976. *Instrumentation for field studies of urban runoff.* Ministry of Environment Research Report 42, Toronto.

Marshall, J.J., 1959. *Relations between water and soil.* Commonwealth Bureau of Soils, Technical Communication 50, Harpenden.

Mather, J.D., 1976. Hydrogeological guidelines for the selection of landfill sites. Appendix I. In: *The licensing of waste disposal sites,* Department of Environment Waste Management Paper No.4, 39-47.

Mather, J.D., 1977. *Attenuation and control of landfill leachates.* 79th Annual Conference Institute of Sewage Works Managers, Torbay, 31st May to 3rd June 1977.

Matthess, G., 1972. Hydrogeological criteria for the self-purification of polluted groundwater. In:

Proceedings of the 24th International Geological Congress, Montreal, Section 11, 296-304.

Mayhead, G.J., 1973. The effect of altitude above sea level on the yield class of Sitka Spruce. *Scottish Forestry,* 27(3), 231-237.

McCulloch, J.S.G., 1969. Director's Review. In: *Institute of Hydrology, Annual Report 1968,* 8-9.

McPherson, M.B., 1969. *Basic information needs in urban hydrology,* American Society of Civil Engineers, New York.

McPherson, M.B., 1972. *Urban runoff.* American Society of Civil Engineers, Office of Water Resources Research, New York.

Metcalf and Eddy Inc., University of Florida, and Water Resources Engineers Inc., 1971. *Environmental Protection Agency storm water management model.* Environmental Protection Agency, 4 Vols., Report Nos. 11024DOC-07/71, 08/71, 09/71, 10/71, Washington.

Ministry of Agriculture, Fisheries and Food, 1976. *Annual Report of the Field Drainage Experimental Unit, 1975.*

Moore, W.L. & Morgan, C.W., 1969. *Effects of watershed changes on streamflow.* University of Texas Press, Austin.

Morgan, W.A., 1962. *Potential evapotranspiration as measured at Valencia observatory over the period August 1952 to February 1962 and a comparison with values as computed by the Penman formula.* Meteorological Service, Dublin, Technical Note 29.

Mosley, M.P., 1975. Channel changes on the River Bollin, Cheshire, 1872-1973. *East Midlands Geographer,* 6, 185-199.

Narayana, V.V.D., and Riley, J.P., 1968. Application of an electronic analogue computer to the evaluation of the runoff characteristics of small watersheds. *International Association Scientific Hydrology Publication* 80, 1, 38-48.

National Soil Survey, Soils Division, 1969. *Ireland General Soil Map.* An Foras Taluntais.

Natural Environment Research Council, 1975a. *Hydrological Research in the United Kingdom (1970-1975).* London.

Natural Environment Research Council, 1975b. *Flood studies report.* 5 Vols, London.

Natural Environment Research Council, 1976a. *Research in geomorphology of water produced landforms.* Natural Environment Research Council Publication Series "B" No.16, London.

Natural Environment Research Council, 1976b. *Report of the Working Party on Hydrology.*

270

Natural Environment Research Council, 1977. *The flood studies report - an opportunity for discussion*. Flood Studies Supplementary Report No.3.

Newson, M.D., 1975. *Flooding and flood hazard in the United Kingdom*. Oxford University Press, Oxford.

Nisbet, J., 1905. *The forester*. Wm Blackwood & Sons, Edinburgh and London.

Nixon, M., 1963. Flood regulation and river training in England and Wales. In: *The conservation of water resources in the U.K.,* Institution Civil Engineers, London, 137-50.

Ochs, H.T., 1975. Modeling of cumulus initiation in METROMEX. *Journal of Applied Meteorology,* 14, 873-882.

Packman, J.C., Lynn, P.P., Beran, M.A., Lowing, M.J. and Kidd, C.H.R., 1976. *The effect of urbanisation on flood estimates*. Institute of Hydrology Urban Drainage Research Note to the National Water Council as WP-HDSS-76/17, and to CIRIA as SG/244/5.

Painter, R.B., Blyth, K., Mosedale J.C. and Kelly, M., 1974. The effect of afforestation on erosion processes and sediment yield. In: *Effects of man on the interface of the hydrological cycle with the physical environment*. International Association of Hydrological Sciences Publication 113, 62-68.

Palmer, C.L., 1950. Pollution effects of stormwater overflows from combined sewers. *Sewerage and Industrial Wastes,* 22, 154-168.

Park, C.C., 1976. Variations and controls of stream channel morphometry. Unpublished Ph.D. thesis, University of Exeter.

Park, C.C., 1977. Some comments on 'The response of natural river channels to urbanisation: two case studies in southeast England'. *Journal of Hydrology,* 32, 193-197.

Penman, H.L., 1950. Evaporation over the British Isles. *Quarterly Journal of the Royal Meteorological Society,* 76, 372-383.

Penman, H.L., 1963a. *Vegetation and hydrology*. Commonwealth Bureau of Soils, Technical Communication 53, Harpenden.

Penman, H.L., 1963b. Woburn irrigation 1951-1959 (1) Purpose, design and weather. *Journal of Agricultural Science,* 58, 343-348.

Penning-Rowsell, E.C. and Parker, D.J., 1974. Improving floodplain development control. *The Planner,* 60, 540-543.

Pereira, H.C., 1973a. Research progress to watershed management. In: *A view from the watershed*. Institute of Hydrology Report 20, 27-36.

Pereira, H.C., 1973b. *Land use and water resources.* Cambridge University Press, London.

Pereira, H.C., Dagg, M. and Hosegood, P.H., 1962. The development of tea estates in tall rain forests: the water balance of both treated and control valleys. *East African Agriculture and Forestry Journal,* 27, 36–40.

Petts, G.E., 1977. The impact of reservoir construction upon stream channels: a case study. In: Gregory, K.J. (Editor), *River channel changes,* Wiley, London, 145–164.

Phillips, A.D.M. and Clout, H.D., 1970. Underdraining in France during the second half of the nineteenth century. *Transactions of the Institute of British Geographers,* 51, 71–94.

Pluhowski, E.J., 1970. Urbanisation and its effect on the temperature of the streams of Long Island, New York. *U.S. Geological Survey Professional Paper* 627-D.

Popov, I.V. and Gavin, Y.S., 1970. Use of aerial photography in evaluating flooding and emptying of river floodplains and the development of floodplain currents. *Soviet Hydrology,* 5, 413–425.

Puertner, P.E., 1976. Urban stormwater detention and flow attenuation. *Public Works,* 52, 83–85.

Quirk, J.P. and Panabokke, C.R., 1962. Incipient failure of soil aggregates. *Journal of Soil Science,* 13, 60–70.

Rakhmanov, V.V., 1966. *Role of the forest in water conservation.* Translation from the Russian, Israel Program for Scientific Translations, Jerusalem.

Rao, R.A., Delleur, J.W. and Sarma, B.S.P., 1972. Conceptual hydrologic models for urbanising basins. *Proceedings of the American Society Civil Engineers, Journal of the Hydraulics Division,* 98, No. HY7, 1205–1220.

Rayment, A.F. and Cooper, D.J., 1968. Drainage of Newfoundland peat soils for agricultural purposes. In: *Proceedings of the Third International Peat Congress,* Quebec, Canada, 345.

Reid, G.K., 1961. *Ecology of inland waters and estuaries.* Reinhold, New York.

Reid, I., 1975. Seasonal variability of rainwater redistribution by field soils. *Journal of Hydrology,* 25, 71–80.

Reynolds, E.R.C., and Henderson, C.S., 1967. Rainfall interception by beech, larch, and Norway spruce. *Forestry,* 40(2), 165–184.

Richards, K.S. and Wood, R., 1976. *Urbanisation, water redistribution and their effect on channel processes.* Paper presented at Institute of British

Geographers Annual Conference, Lanchester Polytechnic, Coventry.

Richardson, S.J., 1976. Effect of artificial weathering
cycles on the structural stability of dispersed
silt soil. *Journal of Soil Science,* 27, 287-294.

Richter, G., 1965. *Bodenerosion schäden und gefährdete
gebiete in der Bundesrepublic Deutschland,* Bad
Godesberg.

Riggs, H.C., 1968. Some statistical tools in hydrology.
In: *Techniques of Water Resources Investigations
of the U.S. Geological Survey,* Book 4, Ch.A1,
Washington D.C.

Rijtema, P.E., 1968. *On the relation between transpiration,
soil and physical properties and crop production
as a basis for water supply plans.* Institut voor
Cultuurtechniek en Waterhuishouding, Technical
Bulletin 58, Wageningen.

Road Research Laboratory, 1963. *A guide for engineers to
the design of storm sewer systems.* Road Note 35,
H.M.S.O., London.

Roberts, M.E. and James, D.B., 1972. Some effects of
forest cover on nutrient cycling and river temperature. In: Taylor, J.A. (Editor), *Research papers in
forest meteorology,* The Cambrian News, Aberystwyth,
100-108.

Rodda, J.C., 1967. The rainfall measurement problem. In:
*Proceedings of the International Association of
Scientific Hydrology General Assembly, Bern,
Publication 78,* 215-231.

Rose, C.W., 1960. Soil detachment caused by rainfall.
Soil Science, 89, 28-35.

Ross, F.F., 1959. The operation of thermal stations in
relation to streams. *Journal and Proceedings of the
Institute of Sewage Purification,* 16, 2-11.

Ross, F.F., 1970. *Warm water discharges into rivers and
the sea.* Annual Conference of Institute of Water
Pollution Control, 8-11 Sept., Blackpool.

Russell, E.W., 1971. Soil structure: its maintenance and
improvement. *Journal of Soil Science,* 22, 137-151.

Rutter, A.J., 1964. Studies in the water relations of
Pinus sylvestris in plantation conditions. II.The
annual cycle of soil moisture change and derived
estimates of evaporation. *Journal of Applied
Ecology,* 1, 29-44.

Ryabchikov, A., 1975. *The changing face of the earth.*
Progress Publishers, Moscow.

Rycroft, D.H., Williams, D.J.A. and Ingram, H.A.P., 1975.
The transmission of water through peats, I, Review,
Journal of Ecology, 63, 535-556. II, Field experi-
ments, *Journal of Ecology,* 63, 557-568.

Sartor, J.D. and Boyd, G.B., 1972. *Water pollution aspects of street surface contaminants*. Environmental Protection Agency Report R2-72-081, Washington D.C.

Sauer, V.B., 1974. *An approach to estimating flood frequency for urban areas in Oklahoma*. U.S. Geological Survey, Water Resources Investigations 23-74, Oklahoma.

Savini, J. and Kammerer, J.C., 1961. *Urban growth and water regimen*. U.S. Geological Survey Water Supply Paper 1591-A.

Schmudde, T.H., 1963. Some aspects of the landforms of the lower Missouri River floodplain. *Annals of the Association of American Geographers,* 53, 60-73.

Schneider, W.A., 1969. Reforestation effects on winter and spring flood peaks in central N.Y. State. In: *Floods and their computation,* 2, International Association Scientific Hydrology UNESCO/WMO, 750-789.

Schumm, S.A., 1954. The relation of drainage basin relief to sediment loss. In: *International Association Scientific Hydrology, Publication 36,* 216-219.

Schumm, S.A., 1960. The shape of alluvial rivers in relation to sediment type. *U.S. Geological Survey Professional Paper 352-B,* 17-30.

Schumm, S.A., 1969. River metamorphosis. *Proceedings of the American Society of Civil Engineers, Journal of the Hydraulics Division,* HY1, 6352, 255-273.

Semonin, G.R. and Changnon, S.A., 1974. METROMEX: Summary of 1971-1972 results. *Bulletin of the American Meteorological Society,* 55, 95-99.

Serr, B.F., 1972. Unusual sediment problems in N. coastal California. In: Shen, H. (Editor), *Sedimentation* (Einstein Symposium Volume). Water Resources Publication, Fort Collins, Colorado.

Slack, J.G., 1977. Nitrate levels in Essex river waters. *Journal of the Institution of Water Engineers and Scientists,* 31, 43-51.

Slaymaker, H.O., 1972. Patterns of present sub-aerial erosion and landforms in mid-Wales. *Transactions of the Institute of British Geographers,* 55, 47-68.

Smith, K., 1968. Some thermal characteristics of two rivers in the Pennine area of northern England. *Journal of Hydrology,* 6, 405-416.

Smith, K., 1972. River water temperatures: an environmental review. *Scottish Geographical Magazine,* 88, 211-220.

Smith, K., 1975. Water temperature variations within a major river system. *Nordic Hydrology,* 6, 155-169.

Smith, K., 1976. Hydrological research in the United Kingdom, 1965-75. *Area,* 8(4), 273-277.

Smith, K. and Lavis, M.E., 1975. Environmental influences

on the temperature of a small upland stream. *Oikos*, 26, 228–236.

Smith, R.E. and Woolhiser, D.A., 1971. Overland flow on an infiltrating surface. *Water Resources Research*, 7, 899–913.

Smith, R.N., 1976. Nutrient budget of the River Main, Co. Antrim. Ministry of Agriculture, Fisheries and Food Technical Bulletin No.32, In: *Agriculture and Water Quality*, 315–339.

Snyder, F.F., 1938. Synthetic unit-graphs. *Transactions of the American Geophysical Union*, 19, 447–454.

Sopper, W. and Lull, H., 1965. *International Symposium on Forest Hydrology*. Pergamon, Oxford.

Steinhardt, R. and Trafford, B.D., 1974. Some effects of sub-surface drainage and ploughing on the structure and compactibility of a clay soil. *Journal of Soil Science*, 25, 138–152.

Stewart, J.B., 1971. The albedo of a pine forest. *Quarterly Journal of the Royal Meteorological Society*, 97, 561–564.

Stoneham, S.M. and Kidd, C.H.R., 1977. *Prediction of run-off volume from fully sewered urban catchments*. Institute of Hydrology Report No.41.

Sturges, D.L., 1968. Hydrologic properties of peat from a Wyoming mountain bog. *Soil Science*, 106, 262–264.

Swartzendruber, D. and Hillel, D., 1975. Infiltration and runoff for small field plots under constant intensity rainfall. *Water Resources Research*, 11, 445–451.

Tallis, J.H., 1973. Studies on southern Pennine peats. Direct observations on peat erosion and peat hydrology at Featherbed Moss, Derbyshire. *Journal of Ecology,* 61, 1–22.

Tamm, C.O., Holman, H., Popovic, B. and Wiklander, G., 1974. Leaching of plant nutrients from soils as a consequence of forestry operations. *Ambio*, 3(6), 211–221.

Thomas, W.L., 1956. *Man's role in changing the face of the earth*. University of Chicago Press.

Thompson, R.W.S., 1954. Stratification and overturn in lakes and reservoirs. *Journal of the Institution of Water Engineers*, 8, 19–36.

Turner, D.J., 1971. Dams in ecology. *Civil Engineering* (American Society Civil Engineers), 41(9), 76–80.

Turner, D.J., 1977. *The safety of the herbicides 2,4-D and 2,4,5-T*. Forestry Commission Bulletin 57, H.M.S.O., London.

Twort, A.C., Hoather, R.C. and Law F.M., 1974. *Water supply*. 2nd Edition, Arnold, London.

UNESCO, 1972. Influence of man on the hydrological cycle: Guidelines to policies for the safe development of land and water resources. In: *Status and trends of research in hydrology 1965-74,*Studies and reports in hydrology 10, UNESCO, 31-70.

UNESCO, 1974. *Hydrological effects of urbanization.* Studies and reports in hydrology 18, UNESCO.

UNESCO, 1975. *Hydrology of marsh ridden areas.* UNESCO-IAHS.

U.S. Army Corps of Engineers, 1971. *Hydrologic Engineering methods for water resources development, Vol. 1, Requirements and general procedures.* Davis, California.

U.S. Department of Agriculture, Soil Conservation Service, undated. *Computer program for project formulation - hydrology.* Soil Conservation Service, Technical Release SCS-TR-20, Washington.

U.S. Department of Agriculture, Soil Conservation Service, 1975. *Urban hydrology for small watersheds.* Soil Conservation Service, Technical Release SCS-TR-55, Washington.

U.S. Federal Water Pollution Control Federation, 1969. *Water pollution aspects of urban runoff.* American Public Works Association, Washington D.C.

Van Post, L. and Granlund, E., 1926. Sodra Sveriges tortvtillgangar. I. *Sveriges Geologiska Undersokning, Serie C,* 335.

Van Sickle, D.R., 1963-4. The effects of urban development on storm runoff. *The Texas Engineer,* 32, No.12, 3-5, and 33, No.1, 2-7.

Vice, R.B., Guy, H.P. and Ferguson, G.E., 1969. *Sediment movement in an area of suburban highway construction, Scott Run Basin, Fairfax County, Virginia, 1961-64.* U.S. Geological Survey Water Supply Paper 1591-E.

Wallace, J.R., 1971. *The effects of land use change on the hydrology of an urban watershed.* Georgia Institute of Technology, Environmental Resources Centre, Report ERC-0871, Atlanta.

Waller, D.H., 1972. Factors that influence variations in the composition of urban surface runoff. *Water Pollution Research in Canada,* 7, 68-95.

Walling, D.E., 1971. Sediment dynamics of small instrumented catchments in south-east Devon. *Transactions of the Devonshire Association,* 103, 147-165.

Walling, D.E., 1974. Suspended sediment and solute yields from a small catchment prior to urbanisation. In: Gregory, K.J. and Walling, D.E. (Editors), *Fluvial processes in instrumented watersheds.* Institute British Geographers Special Publication 6, 169-191.

Walling, D.E., 1977. Physical hydrology. *Progress in Physical Geography,* 1(1), 143-151.

Walling, D.E. and Gregory, K.J., 1970. The measurement of the effects of building construction on drainage basin dynamics. *Journal of Hydrology,* 11, 129-144.

Walling, D.E., and Teed, A., 1971. A simple pumping sampler for research into suspended sediment transport in small catchments. *Journal of Hydrology,* 13, 325-337.

Ward, R.C., 1963. Measuring potential evapotranspiration. *Geography,* 48, 49-55.

Ward, R.C., 1975. *Principles of hydrology.* 2nd Edition, McGraw-Hill, London.

Water Resources Board, 1972. *The hydrogeology of the London Basin.* 3 vols. Water Resources Board, Reading.

Water Resources Board, 1973. *Ninth Annual Report - year ending 30th September 1972.* H.M.S.O., London.

Waterton, T. *et al.*, 1969. The effects of tipped domestic refuse on groundwater quality (Symposium). *Water Treatment and Examination,* 18, 15-69.

Watkins, L.H., 1962. *The design of urban sewer systems.* Road Research Laboratory Technical Paper No.55, H.M.S.O., London.

Weibel, S.R., Anderson, R.J., Woodward, R.L., 1964. Urban land runoff as a factor in stream pollution. *Journal of the Water Pollution Control Federation,* 36, 914-924.

Wellbank, P.J., Gibb, M.J., Taylor, P.J. and Williams, E.D., 1974. Root growth of cereal crops. In: *Rothamsted Experimental Station Annual Report for 1973,* 26-66.

Wesseling, J., 1958. *Hydrology, soil properties, crop growth and land drainage.* Institut voor Cultuurtechniek en Waterhuishouding Technical Bulletin 57, Wageningen.

Whipple, W., 1975. *Urban runoff: quantity and quality,* American Society Civil Engineers, New York, 272 pp.

Whipple, W., Hunter, J.V. and Yu, S.L., 1974. Unrecorded pollution from urban runoff. *Journal of the Water Pollution Control Federation,* 46, 837-885.

Wilcock, D.N., 1977a. Water resource management in Northern Ireland. *Irish Geography,* (in press).

Wilcock, D.N., 1977b. The effects of channel clearance and peat drainage on the water balance of the Glenullin basin, County Londonderry, Northern Ireland. *Proceedings of the Royal Irish Academy,* (in press).

Wilkinson, R., 1956. Quality of rainfall runoff water from a housing estate. *Journal of the Institute Public Health Engineers,* 55, 70-84.

Wilton, B., 1964. Effect of cultivations on the level of the surface of a soil. *Journal of Agricultural Engineering Research,* 9, 214-219.

Wischmeier, W.H. and Smith, D.D., 1958. Rainfall energy

and its relation to soil loss. *Transactions of the American Geophysical Union,* 39, 285-291.

Wischmeier, W.H. and Smith, D.D., 1965. *Predicting rainfall-erosion losses from cropland east of the Rocky Mountains.* U.S. Department of Agriculture, Agricultural Research Service, Agricultural Handbook 282.

Wisler, C.O., and Brater, E.F., 1959. *Hydrology.* Wiley, New York.

Wolman, M.G., 1959. Factors influencing erosion of a cohesive river bank. *American Journal Science,* 257, 206-216.

Wolman, M.G., 1967. A cycle of sedimentation and erosion in urban river channels. *Geografiska Annaler,* 49A, 385-395.

Wolman, M.G., 1975. Erosion in the urban environment. *Bulletin of the International Association Hydrological Sciences,* 20, 117-125.

Wolman, M.G. and Schick, A.P., 1967. Effects of construction on fluvial sediment, urban and suburban areas of Maryland. *Water Resources Research,* 3, 451-464.

Woodward, H.B., 1904. *Water supply of Lincolnshire from underground sources.* Memoirs Geological Survey, U.K.

Woodward, H.B., 1906. The utilisation of old pits and quarries, and of cliffs, for the reception of rubbish. *Journal of the Royal Sanitary Institute,* 27, 467-469.

World Meteorological Organisation (WMO), 1970a. *Urban climates.* WMO Technical Note 108.

World Meteorological Organisation (WMO), 1970b. *Building climatology.* WMO Technical Note 109.

Worthington, P.F., 1977. Permeation properties of the Bunter sandstone of northwest Lancashire, England. *Journal of Hydrology,* 32, 295-303.

Wyatt, R.J., Horton, A. and Kenna, R.J., 1971. Drift-filled channels on the Leicestershire-Lincolnshire border. *Bulletin of the Geological Survey of Great Britain,* 37, 57-81.

Yorke, T.H. and Davis, W.J., 1972. Effects of urbanisation on sediment transport in Bel Pre Creek Basin, Maryland. *U.S. Geological Survey Professional Paper* 750-B, 218-223.

Young, A., 1958. A record of the rate of erosion on Millstone Grit. *Proceedings of the Yorkshire Geological Society,* 31, 49-56.

Young, R.A., and Wiersma, J.L., 1973. The role of rainfall impact in soil detachment and transport. *Water Resources Research,* 9, 1629-1636.